AF658258

Encyclopaedia of
Environment and Development

Encyclopaedia of Environment and Development

VOLUME - II

FALENDRA K. SUDAN

SERIALS PUBLICATIONS
NEW DELHI

Published - 2019

ISBN: 978-81-8387-195-2

Published by

SERIALS PUBLICATIONS
4830/24, Prahlad Street, Ansari Road
Darya Ganj, New Delhi-110002 (India)
Phone : 23245225. Fax : 91-11-23272135
E-mail: serials@satyam.net.in

CONTENTS

PREFACE

Jean-Pierre Amigues and Michel Moreaux's paper titled "Contrasting how to overcome the resource constraints in non-renewable and renewable resource economies: Does the shape of the utility functions matter?" show that optimal research policies are quite different in non renewable and renewable resource economies. We assume first that the utility functions are isoelastic functions and that the welfare is the sum of the instantaneous utilities discounted at some social rate of impatience, and second that the production technologies are Leontiev production functions. We demonstrate that in renewable resource economies the optimal research policy is what we call a least effort policy whereas in non renewable economies the research effort is depending upon the previously accumulated technological knowledge and the available stock of resource.

In the paper titled "Common property resources management in India: Some review findings" by Naresh Chandra Sahu and Binayak Rath, a debate on the management of common property resources (CPRs) has been discussed in the light of Hardin's Tragedy of Commons. The emergence of participatory forest resources management approach in India has also been analyzed. The issues and challenges in the management of the forest resources have been highlighted in the context of adoption of participatory management approach undertaken by the government of India since 1988. The literature on degradation and decline of forests has been reviewed in order to identify the possible principal factors responsible for the same. The paper also brings out the major determinants of CPR management institutions. The issue of efficiency and equity pertaining to CPR management institutions has been critically discussed. The role of socio-economic heterogeneity on the efficacy of CPR has been critically. The paper also focuses on the literature unearthing the relationship between the CPR institutions and stress migration.

Zakir Husain and Dr. Rabindra Nāth Bhattacharya's paper titled "Privatizing the commons: A critical review of the property rights paradigm" emphasizes that the commons problem emerges because the existing property rights structure cannot prevent the dissipation of the benefits of environmentally conscious behaviour to free riders. The property rights school, based on the Coasian concept of externality, argue that privatisation of the resource will resolve the crisis in such situations. Biologists (Wynne-Edwards Ardrey) and psychologists (Crowe, Edney) have also supported this view. In the present paper, we have critically examined their arguments. We have pointed out that the property rights school fails to consider the costs of privatisation and the fact that the prevailing system of property rights depends upon the surplus being generated. They also ignore the considerable body of literature documenting the depletion of privatised resources. The real cause of the crisis, we argue, is not the system of property rights per se, but in the delineation of the rights over possible income flows from the resource. Proper attenuation – the restriction of the rights to use of the resource – can resolve the commons dilemma. This view accepts community management as an important and distinct form of resource regimes that simultaneously ensures inter-generation and intra-generational equity.

The study titled "Does socio-economic heterogeneity matter in the community based forest resource management? An empirical investigation from Orissa" by Naresh Chandra Sahu and Binayak Rath attempts to examine the impact of socioeconomic heterogeneity at the household levels on the probability of participating in the management of forest protection committees. The results of our empirical investigation using Logit regression model suggest that the socio-economic heterogeneity has an ambiguous impact on the probability of the participation and formation of community based forest resources management in Orissa. Moreover, the interesting finding of the study is that cast heterogeneity does not have any significant relation with the collective action like participating in forest protection committees (FPC) activities contrary to the views of some scholars. However, the education level of households has significant effect on the probability of participating in the community based forest resources management. But the land size and income inequality may hinder probability of joining the FPCs activities and programmes in the state.

Kuntala Lahiri-Dutt and Gopa Samanta's paper titled "Uncertain livelihoods: Survival strategies in charland environments in India" reveals that charlands are river islands found on riverine plains and characterized by vulnerable and difficult living conditions and extreme resource constraints. The charlands may often be temporary pieces of land, their existence dependent on the rise and fall of the river. The inhabitants of chars have been dispossessed and are the poorest and most disadvantaged members of society; often migrants from other parts of the country or those who have illegally crossed the international border. In Southern Bengal, the chouras or char inhabitants have adapted to this marginal environment and have developed certain livelihood strategies to ensure their survival. This paper reports on the livelihood strategies of these charland inhabitants based on a study of women and men living in the highly dynamic charland environments of the Damodar River in Southern Bengal, India.

The paper titled "Traditional environmental knowledge of Bawaliis: A case study of the Sundarbans mangrove forest" by Md. Mustafa Kamal Akand reveals that the Sundarbans Mangrove forest area is largely inhabited by the occupational groups of people who are generally engaged with gathering of wood, leaves, honey, fish and garnering of fry and the like. These forest people inhabit a different setting of ecosystem with their traditional means of interaction. They rely heavily on the natural resources of surrounding forest and coastal river for their livelihood, which is regarded to be a risky and unstable habitat. Unlike the mainlanders, the pattern of subsistence of that forest people is submerged with several uncertainty and vulnerability, as they always encounter with the wild beasts and other environmental adversity. In this mangrove forest, like other inhabitants the Bawali groups developed a unique and sustainable system of interaction with their ecosystem and maintaining a state of equilibrium between nature and mankind. This paper focuses on the indigenous pattern of environmental knowledge regarding their subsistence and thus explores to know as to how they are utilizing traditional knowledge to cope with environmental risk and danger. This paper concludes with some policy recommendations that may be considered by the planners and implementers.

While investigating the historical perspective of joint forest management programme (JFMP), the paper titled "Historical relevance of joint forest management programme and the key element of its sustainability: Evidence from western Midnapore division of West Bengal" by Debnarayan Sarker and Nimai Das observes that the resistance movement of forest communities in western Midnapore division in West Bengal, which acted as key precursor to joint forest management (JFM) programme in India through a June 1990 Ministry of Environment and Forest (MOEF) circular based upon, to a large extent, the successful experience of joint management in Arabari hills, under this division, have been mobilized by the poor forest communities from long past for their community right on their forest resources in connection with the immediate survival needs. The study of existing four-forest protection committees (FPCs) of this area also confirms that this immediate survival needs generating mainly from non-timber forest products' (NTFPs) income of FPC members are the key element for the sustainability of JFM.

In the paper titled "Institutional change and common property resource management in Shivaliks, Jammu and Kashmir" by Falendra K. Sudan, an attempt has been made to examine common property resource (CPR) management and cost and benefit sharing mechanisms in Shivaliks, Jammu and Kashmir in the context of Integrated Watershed Development Project (IWDP), Hills-II. The project in association with VDCs has created assets through closing degraded forestland, village common lands (VCLs) and constructing water-harvesting structures. Recently, election of panchayats has been held, which would also pave the way in motivating the beneficiaries in cost sharing. Over the period, the proportion of the costs shared by the beneficiaries has improved and on average, it was reportedly more than 25 per cent. The benefits from assets created have started accruing to the population of adjacent areas. To maintain these assets and ensure benefits sharing, the project in association with VDCs has formed user groups. The constitution, number of members and the method of benefit sharing has been different for each user groups and devised by the group for its sustenance and establishing linkage with VDC. The user participation in pre-project planning such as information gathering, identification of the resources to be protected and the targeted groups to be involved, and the design of the regulations related to monitoring and enforcement has substantial bearing on the sustainability of participatory watershed development. Thus, it is suggested to recognize the poor villagers not only as beneficiaries of the participatory resource management, but to assign them the status of stakeholders in real practice and provide them the opportunities to participate not only in conservation and protection phase, but equally in pre-project phase also. Wider linkages are essential between local groups and development institutions, which will ensure institutional sustainability. Among others, links are needed with credit institutions, with agencies providing development investments and with institutions of local government, particularly the panchayats. To achieve this, project promoted structures have to incorporate the concerns and leadership of the panchayats, without losing the advantages of local user control and management of CPRs.

The paper titled "Common property resources and sustainable development: A perspective" by M. Sabesh Manikandan, K. Raja Mohan Rao and S. Nehru reveals that the socio-economic development of man depends on the environment and this development influences the environment. Thus environment and development are interdependent. Now-a-days 'development' has become a comprehensive word carrying such wide as well as specific meanings that both the common man and the specialist can use it in their own contexts. In fact, economic development is closely linked to the development of resources. These resources can be classified into three categories, viz., public goods, semi-public goods and private goods. The term 'Common Property Resources' (CPRs) is broadly defined as natural resources in which a group of people has common user rights (not necessarily ownership rights). These include natural forests;

village lands, grazing lands, streams, rivers, groundwater and oceans. These may also include man-made resources like irrigation tanks, community wells and village roads. Sustainable Development is about enhancing human well being through time. One should have the ability and opportunity to shape one's life, which increase with better health, education and material comforts, enhancing the portfolio of assets. The World Commission on Environment and development defined sustainable development as "Progress that meets the needs of the present without compromising the ability of the future generations to meet their own need". In this regard the present paper presents a new perspective on sustainable development.

Yogesh Ranjit's paper titled "Participatory forest management: A survey of review of literature" reveals that since the last three decades, participatory forestry is being increasingly considered as a chief strategy for conservation, management and sustainable use of forests. This approach is being widely promoted in various forms with varying degrees of sharing of rights and responsibilities in forest management and forest-benefits between local communities and forest department. Besides, a number of civil societies, non-governmental organization (NGOs) and international donor agencies have come into existence to influence, facilitate and promote the sustainable use and management of forests. In this regard, the article seeks to provide an overview of review of literature on Participatory Forest Management (PFM) in Global context especially focusing on the experience of India and Nepal. Syed Ajijur Rahman and Khandaker Mursheda Farhana's paper titled "Economic evaluation of multistrata agroforestry system practiced by traditional cultivators in Northern Bangladesh" reveals that agroforestry is an important category of agriculture that has the potential to provide farmers, communities, and society-at-large with a wide array of forest-related goods and services.T his paper evaluates the selected multistrata agroforestry model in Northern Bangladesh, comparing to traditional system of monoculture. Research data were collected by the means of FGD, household structured interviews and indepth case study. Data were analyzed through quantitative economic methods and models. Research indicates multistrata agroforestry gives positive and much higher net present value (NPV) than traditional monoculture in both scenarios with and without the inclusion of family labor costs and it is more profitable and less risky compared to other agricultural options. In their another paper titled "Cost-benefit analysis of agriculture versus agroforestry system in the high barind tract of Northern Bangladesh" reveals that traditional agriculture in Northern Bangladesh is unsustainable and keeps farmers in a vicious circle of poverty. Agroforestry is a promising alternative. This paper assess the benefit-cost of traditional agriculture and agroforestry in the High Barind

Tract of Northern Bangladesh. Both quantitative and qualitative methods are used for analyze data. Cost-benefit analysis is applied to assess the quantitative data. Results indicate agroforestry may not only be good for (agro) biodiversity and sustainability, but may also have quite good economic rates of return. The agroforestry system gives positive and much higher net present value (NPV) than that of agriculture. The financial analysis also indicates that agroforestry system is less risky compared to many agricultural projects. Appropriate strategies leading to right choice of species i.e., indigenous and fast growing species; optimal species combinations; solving land tenure problems and imparting training to farmers could promote agroforestry system and enhance the farm income.

V. Anitha, P. K. Muraleedharan and A. S. Binilkumar's paper titled "Human dimensions in protected in protected area management: Tracing synergies of conservation and livelihood" emphasizes that forest landscapes rich in biodiversity are fast shrinking having various social, economic and ecological effects, as many people depend on these resources for their sustenance. Underlying causes include issues such as poverty, unequal land ownership, social inequality, population growth, forest dependencies and other commercial interests, which is a cause of concern globally. The demarcation and establishment of Protected Areas with varied status of protection has been found useful to some extent for conservation of biodiversity while on the other hand such propositions have created new threats in the management of the natural resources as the local communities look upon these as government-imposed restrictions on their livelihood systems. Thus, the synergy between the dependent communities and the nature has been disturbed in the process of development. Considering the diversity in the landscape and the livelihood issues involved, the study carried out in two Wildlife Sanctuaries in Kerala-Peechi-Vazhani and Chimmoni- is an attempt to understand the human dynamics involved in Protected Area Management and its associated problems/issues. The paper explores the need for a new landscape-livelihood approach towards Protected Area management in order to understand the intricate linkages.

Bamwerinde W. et. al. in their paper titled "The puzzle of idle land in the densely populated Kigezi highlands of southestern Uganda" characterizes land use at the household level and identify two types of long-term uncultivated lands – those that are intentional fallows and those that are neglected or abandoned. We then use a multinominal logit model to examine the determinants of plot abandonment and long fallows in order to propose policy interventions that lead to optimal and sustainable management of land use systems in Kigezi highlands. Household factors such as age, and formal education positively influenced farmers' decision to abandon plots.. However, farm size and household type had no significant influence on abandonment of plots. Plot variables such as slope and distance between the homestead and the plot had the expected positive signs while soil fertility had a significant negative sign as predicted. From the analysis different typologies of uncultivated lands were defined depending on their inherent characteristics and distance to the homestead. Farmers then provided technological and policy options as to how these types of plots could be made more productive.

Patricia M. Makepe's paper titled "Population density, size of operation and collective action for monitoring and investment in rural infrastructure: Implications for policy" reveals that the provision of adequate rural infrastructure in developing countries is critical for enhancing the performance of income generating activities and the improvement of rural livelihoods. In Botswana cattle rearing is still the most dominant income generating activity for the rural population. For cattle rearing most critical type of infrastructure is the borehole (a deep well or water point) because it enables a cattle farmer to utilize underground water for purposes of watering cattle and a borehole also guarantees access and control to the grazing resource within the vicinity of the borehole. Communities in remote areas usually have to make the most investment in infrastructure. It is common for individuals to get together as a group and act collectively and pool their resources to make the much need investment in developing the borehole and also monitor its use. The findings of the study indicate that collective action for making investments in rural infrastructure in remote areas is important. Understanding the relationship between population density, size of operation and collective action for making investments, and monitoring rural infrastructure is critical as it can help policy makers in identifying ways in which collective action can be enhanced where it is not functioning properly. The paper contributes to the search for appropriate community based resource management strategies for improving rural livelihoods.

The paper titled "Land use pattern and ecological implications: A comparative study of Uttar Pradesh and Madhya Pradesh" by Gyan Prakash and Ram Kumar Jha reveals that land resource is important natural wealth of the nation and its proper utilization is a matter of concern. Its improper use leads to progressive deterioration and create ecological imbalances. The changes in land use pattern affect eco system in terms of vegetation, local climate, land quality, quality of life and water table. An attempt has been made in this paper to examine comparative analysis of the land use pattern and its ecological implications by using the land use changes in Madhya Pradesh and Uttar Pradesh for the period from 1990-91 to 2001-02. The study reveals that the substantial land shifts have taken place from the undesirable ecological sector (Barren land) E_2 to other sectors across the states. This trend is highly appreciable, favourable and positive but the depletion of desirable ecological sector (forest + Permanent Pasture + Miscellaneous trees and groves) E_1 has been observed across the states (Madhya Pradesh and Uttar Pradesh). It is a serious concern for economic growth. It seems, a part of this sectoral land use has shifted towards the non-agricultural sector due to urbanization and industrialization besides these some part of lands have shifted for the development of rural infrastructure etc. In the light of present analysis of land use, government should formulate a policy so that economic growth process may make least harm to environment.

LIST OF CONTRIBUTORS

Jean-Pierre Amigues, INRA and University of Toulouse 1 (IDEI and INRA-LERNA), 21 Allée de Brienne - 31000 Toulouse, France.

Michel Moreaux, Université de Toulouse I, (IUF, IDEI and INRA-LERNA), 21 allée de Brienne - 31000 Toulouse, France.

Naresh Chandra Sahu, Research Scholar of Economics, HSS Department, Indian Institute of Technology, Kanpur -208016, U.P., India.

Binayak Rath, Professor of Economics, HSS Department, Indian Institute of Technology, Kanpur -208016, U.P., India.

Zakir Husain, Senior Lecturer, Department. of Economics, Rabindra Bharati University, West Bengal, India.

Rabindra Nath Bhattacharya, Professor, Department of Economics, Kalyani University, West Bengal, India.

Kuntala Lahiri-Dutt, Research Fellow, RMAP, Research School of Pacific and Asian, The Australian National University, ACT 0200, Australia.

Gopa Samanta, Lecturer in Geography, Lecturer in Geography, Mankar College, Mankar, Burdwan 713144, India.

Md. Mustafa Kamal Akand, Assistant Professor, Department of Anthropology, Rajshahi University, Bangladesh.

Debnarayan Sarker, Professor, Centre for Economic Studies, Presidency College, Kolkata, West Bengal, India.

Nimai Das, Research Fellow, Centre for Economic Studies, Presidency College, Kolkata, West Bengal, India.

Falendra K. Sudan, Reader, Department of Economics, University of Jammu, Jammu, Jammu and Kashmir, 180 006, India.

M. Sabesh Manikandan, Research Scholar, Department of Economics, Gandhigram Rural Institute, (Deemed University), Gandhigram, Tamil Nadu, India.

K. Raja Mohan Rao, Reader, Department of Economics, Gandhigram Rural Institute, (Deemed University), Gandhigram, Tamil Nadu, India.

S. Nehru, Senior Lecturer, Department of Economics, Gandhigram Rural Institute, (Deemed University), Gandhigram, Tamil Nadu, India.

Yogesh Ranjit, Ph.D. Research Scholar, Department of Economics, University of Jammu, Jammu, J&K, India - 180 006 and Senior Lecturer, Patan Multiple campus, Department of Economics, Patan Dhoka, Tribhuvan University, Kathmandu, Nepal.

Syed Ajijur Rahman, Research Fellow, Department of Sociology, University of Rajshahi, Rajshahi 6205, Bangladesh.

Khandaker Mursheda Farhana, Research Fellow, Department of Sociology, University of Rajshahi, Rajshahi 6205, Bangladesh.

V. Anitha, Forestry and Human Dimensions Programme Division, Kerala Forest Research Institute, Peechi, Thrissur - 680 653, Kerala, India.

P. K. Muraleedharan, Forestry and Human Dimensions Programme Division, Kerala Forest Research Institute, Peechi, Thrissur - 680 653, Kerala, India.

A. S. Binilkumar, Forestry and Human Dimensions Programme Division, Kerala Forest Research Institute, Peechi, Thrissur - 680 653, Kerala, India.

Bamwerinde W, Forestry Research Resources Institute of the National Agricultural Research Organization, P.O. Box 311, Kabale, Uganda.

B. Bashaasha, Department of Agricultural Economics and Agribusiness, Faculty of Agriculture, Makerere University, P.O. Box 7062, Kampala.

W. Ssembajjwe, Department of Forest Management, Faculty of Forestry and Nature Conservation, Makerere University, P.O. Box 7062, Kampala.

F. Place, World Agroforestry Centre, P. O. Box 3067700100, Nairobi, Kenya.

Patricia M. Makepe, Lecturer, Department of Economics, University of Botswana, Private Bag UB 00705, Gaborone, Botswana.

Gyan Prakash, Reader, School of Economics, Devi Ahilya University, Indore, India.

Ram Kumar Jha, Reader, School of Economics, Devi Ahilya University, Indore, India.

21

CONTRASTING HOW TO OVERCOME THE RESOURCE CONSTRAINTS IN NON-RENEWABLE AND RENEWABLE RESOURCE ECONOMIES – DOES THE SHAPE OF THE UTILITY FUNCTION MATTER

Jean-Pierre Amigues[1] *and Michel Moreaux*[2]

INTRODUCTION

For a long time the main source of primary energy has been the sun energy and its direct derivatives, hydropower and windpower, whereas the less direct derivatives, biomass, animals and humans, was the main source of secondary energy. Although it would be hazardous to claim that the technological progress played no part in the dynamics of these traditional societies, it can be maintained that the speed at which their technology evolved was low, almost zero in a lot of them, even if some of these societies developed a strong corpus of technological knowledge and often very sophisticated devices in order to survive. Those are the type of societies which have been broadly qualified by Wrigley in a provocative essay as *organic* societies[3]. Things begun to change drastically once the occidental societies found the means to exploit on a large scale the stocks of peat and coal, then the stocks of oil and gaz, and finally the stocks of uranium and thorium. In these societies, qualified as *mineral based energy economies* by Wrigley[4], the technological and scientific progress seems more firmly grounded, at least over the two and a half centuries of observations since the beginning of the co-called first industrial revolution.

We suggested in an earlier paper[5] that this strong difference in technological advances could be the byproduct of the type of resource these societies are exploiting, that is a byproduct of a fundamental difference in research incentives within the two types of societies, leaving aside the basic problem of the big-bang switch from the first to the second type of economy. In order to simplify matters we assumed in this earlier essay that the objective function of the society was to maximize the discounted sum of the instantaneous consumption rates, and we showed that, according to the resource is a renewable resource or a non renewable resource, the optimal policy is either an active policy or a passive policy.

Active and passive policies are more easily defined if we assume that the technology is a Leontiev technology, at least for the renewable resource economy as we shall see in section 4 below. What we show in the present paper is that the drastic difference in the optimal research policies is robust to a more general specification of the objective function of the society.

THE MODEL

We consider economies in which a Hicksian aggregate consumption good can be obtained from the joint use of labor and some natural resource, either a non renewable exhaustible resource or a renewable resource. In order to simplify we will assume that only one type of resource is available and we compare what is happening in the economy endowed with the non renewable resource with what is happening in the economy endowed with the renewable resource. Whatever the natural resource with which the economy is endowed, we assume that the some amount of labor is (inelastically) supplied. Without loss of generality we may assume that the amount of available labor is equal to 1.

Let $i \in \{e, r\}$, e for exhaustible and r for renewable, be the indices of each type of economy, q_{it} be the instantaneous production rate of the consumption good, l_{it} be the instantaneous labor input in the consumption good industry, s_t be the instantaneous rate of extraction of the non renewable resource and r_t be the instantaneous rate of use of the renewable resource. Labor and resources are strictly complementary in each economy so that: in the non renewable resource economy:

$$c_t = q_{et} = \min\{A\lambda_{et}, B_{et}s_t\}, \qquad (2.1.e)$$

the renewable resource economy:

$$c_t = q_{rt} = \min\{A\lambda_{rt}, B_{rt}r_t\} \qquad (2.1.r)$$

The technical coefficient of labor A_i, $i \in \{e, r\}$, are constant over time, but the technical coefficients of the resources B_{it} can be improved through dedicated research efforts n_{it}. B_{it} may be understood as a stock of technical knowledge which can be accumulated. We assume that the instantaneous increases of these stocks of knowledge are proportional to the labor input in the research sector n_{it} and to the stock of knowledge previously accumulated:

$$\dot{B}_{it} = b_i n_{it} B_{it}, \quad t \geq 0, \quad b_i > 0. \quad i \in \{e. r\}. \qquad (2.2.i)$$

The initial technical levels are given and denoted by b_t^0, $i \in \{e, r\}$.

Labour can be instantaneously transferred at zero cost from the consumption good industry to the research sector and vice versa. Thus the only constraints on the use of labor are:

$$1 - \lambda_{it} - n_{it} = 0.\ \lambda_{it} \geq 0 \quad \text{and} \quad n_{it} \geq 0,\ t \geq 0,\ i \in \{e, r\}. \tag{2.3.i}$$

In the non renewable resource economy there exists initially a finite stock of the resource denoted by S^0. There is no cost of extraction. Alternatively the production function (2.1.*e*) may be understood as the production function of an integrated process the inputs of which are labor, resource and technical knowledge capital and the output is the consumption good. The dynamics of the resource stock S_t is driven by:

$$\dot{S}_t = -s_t,\ S_t \geq 0 \quad \text{and} \quad s_t \geq 0,\ t \geq 0. \tag{2.4.e}$$

In the renewable resource economy there exists a permanent natural flow of resource, assumed to be constant and denoted by $\bar{r}$. The resource cannot be stored, hence this part $\bar{r} - r_t$ which is not used immediately, is definitively lost. As for the case of the non renewable resource economy we assume zero extraction cost, i.e. the production function (2.1.*r*) may be seen as the production function of an integrated process. The constraints on r_t are:

$$\bar{r} - r_t \geq 0 \quad \text{and} \quad r_t \geq 0. \tag{2.4.r}$$

Note that under the assumption of strict factor complementarity once the stock of knowledge B_{rt} is equal to $\bar{B}_r \equiv A_r \bar{r}^{-1}$, then all the available labor can be allocated to the consumption good production sector since then the resource constraint disappears. Thus in order there exists a true problem we have to assume that $B_r^0 < \bar{B}_r$.

The instantaneous utility function is the standard isoelastic utility function:

$$u(c_t) = (1 - \varepsilon)^{-1} c_t^{1-\varepsilon},\ \varepsilon > 0 \quad \text{and} \quad \varepsilon \neq 1. \tag{2.5}$$

The social welfare $W(C)$ associated to any consumption trajectory $C = \{c_t,\ t \geq 0\}$, is equal to the sum of the instantaneous utilities discounted at some constant rate $\rho > 0$:

$$W(C) = (1 - \varepsilon)^{-1} \int_0^{\infty} c_t^{1-\varepsilon} e^{-\rho t}\, dt. \tag{2.6}$$

Since $c_t \leq A_i,\ i \in \{e, r\}$, then W is well defined for any feasible path C. We denote by $W_t(C)$ the present value at time 0 of the tail of the consumption path C beginning at time t:

$$W_t(C) = (1 - \varepsilon)^{-1} \int_t^{\infty} c_\tau^{1-\varepsilon} e^{-\rho\tau}\, d\tau. \tag{2.7}$$

and by $V_t(C)$ its value at time t:

$$V_t(C) = e^{\rho t} W_t(C) = (1 - \varepsilon)^{-1} \int_t^{\infty} c_\tau^{1-\varepsilon} e^{-\rho(\tau - t)}\, d\tau. \tag{2.8}$$

Last we denote by $MW(C)$ the value of C in terms of discounted marginal utility:

$$MW(C) = \int_t^{\infty} u'(c_t)\, e^{-\rho t}\, dt \tag{2.9}$$

by $MW_t(C)$ the value of the tail beginning at t:

$$MW_t(C) = \int_t^{\infty} u'(c_\tau)\, e^{-\rho t}\, d\tau, \tag{2.10}$$

and last by $MV_t(C)$ its value at time t:

$$MV_t(C) = e^{\rho t} MW_t(C) = \int_0^{\infty} u'(c_\tau) c_\tau\, e^{-\rho(\tau - t)}\, d\tau. \tag{2.11}$$

In each type of economy the benevolent social planner maximizes $W(C)$ under the specific constraints generated by the type of resource available in the economy.

Note that any optimal program has to be efficient. Hence we may restrict the set of admissible programs to the set of efficient programs.

In the exhaustible resource economy, the efficient programs are the programs for which $A_e \lambda_{et} = B_{et} s_t$ so that:

$$s_t = A_e B_{et}^{-1} \lambda_{et}, \tag{2.12.e}$$

While in the renewable resource economy we must have $A\lambda_{rt} = B_{rt} r_t$ so that:

$$r_t = A_r B_{rt}^{-1} rt. \tag{2.12.r}$$

In the non-renewable resource economy the social planner problem $(P.E)$ is:

$$(P.E)\quad \text{Max}_{\{(\lambda et,\, net),\, t \geq 0\}} (1 - \varepsilon) \int_0^{\infty} (A_e \lambda_{et})^{1-\varepsilon} e^{-\rho} dt$$

$$\bar{r} = -A_e B_{et}^{-1} \lambda_{et}, \; S_t \geq 0, \; t \geq 0 \quad S_0 = S^0 \quad \text{given}$$

$$\dot{B}_{et} = b_e n_{et} B_{et}, \; t \geq 0 \text{ and } \; B_{e0} = B_e^0 \quad \text{given}$$

$$1 - \lambda_{et} - n_{et}, \; \lambda_{et} \geq 0 \; \text{ and } \; n_{et} \geq 0$$

In the renewable economy the social planner problem $(P.R)$ is:

$$(P.R)\quad \text{Max}_{\{(\lambda rt,\, nrt),\, t \geq 0\}} (1 - \varepsilon) \int_0^{\infty} (A_r \lambda_{rt})^{1-\varepsilon} e^{-\rho t} dt$$

$$\dot{S}_t = -A_e B_{et}^{-1} \lambda_{et}, \; S_t \geq 0, \; t \geq 0 \quad S_0 = S^0 \quad \text{given}$$

$$\dot{B}_{rt} = b_r n_{rt} B_{rt}, \; t \geq 0 \text{ and } \; B_{r0} = B_r^0 < \bar{B}_r, \quad \text{given}$$

$$1 - \lambda_{rt} - n_{rt}, \; \lambda_{rt} \geq 0 \; \text{ and } \; n_{rt} \geq 0$$

OPTIMAL DEVELOPMENT IN NON-RENEWABLE RESOURCE ECONOMY

Given that A_e is constant and that λ_{et} is bounded from above by 1, under the strict complementarity assumption in the consumption good sector, what can be expected at best is that the society will choose to invest in technical knowledge capital at a sufficiently high rate to compensate for the progressive exhaustion of the resource. As we shall show this is precisely the case iff the productivity in the research sector, b_e, is sufficiently high relative to the social rate of discount ρ. If this is the case, then there exists an optimal stationary state in which the investment in technical knowledge capital just compensate for the exhaustion of the resource. But if the productivity in the research effort is too low, then the society is thrown down into an infernal trajectory along which the consumption rate is decreasing asymptotically to zero while more and more labor must be devoted to the research sector in order to slow down the fall into the hell (zero consumption rate).

Main Characteristics of the Optimal Plans

Let LE_t be the Lagrangian of the program $(P.E)$:[6]:

$$LE_t = (1-\varepsilon)\,(A_e\lambda_{et})^{1-\varepsilon} e^{-\rho t} - \lambda_t A_e B_{et}^{-1}\lambda_{et} + \mu_{et} b_e n_{et} B_{et}$$
$$+\ \omega_{et}[1-\lambda_{et}-n_{et}] + \gamma_{et,t}\lambda_{e,t} + \gamma_{en,t} n_{et}.$$

The F.O.C's are:

$$\partial LE_t/\partial\lambda_{et} = 0$$

$$\Leftrightarrow \quad A_e u'(c_t)e^{-\rho t} = A_e c_t^{-\varepsilon} e^{-\rho t} = A_e^{1-\varepsilon}\lambda_{et}^{-\varepsilon} e^{-\rho t} = \lambda A_e B_{et}^{-1} + \omega_{et} - \gamma_{e\lambda,t} \qquad (3.1)$$

together with the complementary slackness conditions:

$$\partial LE_t/\partial n_{et} = 0 \quad \Leftrightarrow \quad \mu_{et} b_e B_{et} = \omega_{et} - \gamma_{en,t} \qquad (3.2)$$

together with the complementary slackness conditions:

$$\omega_{et} \geq 0 \quad \text{and} \quad \omega_{et}[1-\lambda_{et}-n_{et}] = 0 \qquad (3.3)$$

$$\gamma_{el,t} \geq 0 \quad \text{and} \quad \gamma_{el,t}\lambda_{e,t} = 0 \qquad (3.4)$$

$$\gamma_{en,t} \geq 0 \quad \text{and} \quad \gamma_{en,t} n_{e,t} = 0. \qquad (3.5)$$

The dynamics of the costate variables must satisfy:

$$\dot{\lambda}_t = -\partial LE_t/\partial S_t \Leftrightarrow \dot{\lambda}_t = 0 \quad \Leftrightarrow \quad \text{lt} = 1 \qquad (3.6)$$

$$\dot{\mu}_{et} = -\partial LE_t/\partial B_{et} \Leftrightarrow \dot{\mu}_{et} = -\lambda A_e B_{et}^{-2}\lambda_{et} - \mu_{et} b_e n_{et}. \qquad (3.7)$$

Last the transversality conditions at infinity are:

$$\lim_{t\uparrow\infty} \lambda S_t = \lambda \lim_{t\uparrow\infty} S_t = 0 \quad \text{and} \quad \lim_{t\uparrow\infty} \mu_{et} B_{et} = 0. \qquad (3.8)$$

Note that, since B_{et} is positive and non decreasing, this last condition implies that $\lim_{t\uparrow\infty}\mu_{et} = 0$, that is the imputed value of the stock if technical knowledge must go down to 0 at infinity. Let us first show that, in the present setting, the imputed value of the resource stock S_t must be equal to the imputed value of the stock of capital knowledge B_{et}.

Proposition 1: In the non renewable resource economy, along an optimal path, at each point of time, the imputed values of the resource stock and the imputed value of the technical knowledge stock must be the same:

$$\lambda S_t = \mu_{et} B_{et}, \; t \geq 0. \tag{3.9}$$

Proof: Let h_t be the difference between the two imputed values:

$$h_t = \lambda S_t = \mu_{et} B_{et}, \; t \geq 0.$$

Then:

$$\dot{h}_t = \lambda \dot{S}_t - \dot{\mu}_{et} B_{et} - \mu_t \dot{B}_{et}.$$

In this expression let us substitute for $\dot{S}_t$ its value $-A_e B_{et}^{-1} \lambda_{et} = -A_e B_{et}^{-1}[1 - n_{et}]$, for $\dot{\mu}_{et}$ its value (3.7) taking into account that $\lambda_{et} = 1 - n_{et}$, and for $\dot{B}_{et}$ its value $b_e n_{et} B_{et}$. We get:

$$\dot{h}_t = 0 \Rightarrow h_t = h_0, \; t \geq 0.$$

By the transversality conditions (3.8), we must have $\lim_{t\uparrow\infty} h_t = 0$, hence $h_0 = 0$ that is $h_t = 0$, $t \geq 0$, hence ▪(3.9).

Now let us remark that, given the utility function, we must have $\lambda_{et} > 0$, $t \geq 0$, along any optimal path. Thus multiplying the both sides of (3.1) by λ_{et}, we get:

$$u'(c_t)e^{-\rho t}c_t = \lambda s_t + \omega_{et} \lambda_{et}$$

Integrating over we obtain:

$$MW_t(C) = \lambda S_t + \int_t^{\infty} \omega_{e\tau} \lambda_{e\tau} \, d\tau,$$

or, given (3.9):

$$MW_t(C) = \mu_{et} B_{et} + \int_t^{\infty} \omega_{e\tau} \lambda_{e\tau} \, d\tau.$$

Proposition 2: In the non-renewable resource economy there exist two ways to decompose the value, in terms of discounted marginal utility, of any optimal consumption plan C:

$$MW_t(C) = \lambda S_t + \int_t^{\infty} \omega_{e\tau} \lambda_{e\tau} \, d\tau = \mu_{et} B_{et} + \int_t^{\infty} \omega_{e\tau} \lambda_{e\tau} \, d\tau. \tag{3.10}$$

Stationary Paths and Optimal Stationary Paths

Let us define a balanced or a regular path as any trajectory of the economy along which the sectoral employment is constant: $\lambda_{et} = \lambda_e$ and $n_{et} = n_e \leq 1 - \lambda_e$. Given that A is constant and that the factors are strictly complementary in the consumption good industry, such states are necessarily stationary states.

We first show that at any instant, whatever the resource capital and the technical knowledge capital with which the economy is endowed at this point of time, the society could choose to go along some efficient balanced path. We next show that there exists one and only one optimal steady state of the economy provided that the productivity of factors in the research sector, b_e, is higher than the rate of impatience or social discount rate ρ.

Regular Paths

We determine how the economy could switch to a stationary path at any point of time t_0. Since the research effort is constant, then the technical knowledge capital must grow exponentially:

$$B_{et} = B_{et_0} e^{b_e n_e (t - t_0)}, \ t = t_0.$$

So that, under the efficiency condition (2.12.*e*), the resource extraction rate must decrease at the same rate:

$$s_t = A_e B_{et_0}^{-1} \lambda_e e^{-b_e n_e (t - t_0)}, \ t \geq t_0.$$

Along such a path, for a sectoral employment $(\hat{\lambda}_e, n_e)$, the amount of resource used in the consumption good sector over the time interval $[t_0, t]$, denoted by $\overline{S}_{t_0 t}(\lambda_e, n_e)$, is given by:

$$\overline{S}_{t_0 t}(\lambda_e, n_e) = A_e B_{et_0}^{-1} \lambda_e \int_{t_0}^{\infty} e^{-b_e n_e (\tau - t_0)} d\tau = A_e B_{et_0}^{-1} \lambda_e (b_e n_e)^{-1} [1 - e^{-b_e n_e (t - t_0)}].$$

Thus the society must have to its disposal at the time an amount of non-renewable resource equal to:

$$\lim_{t \uparrow \infty} \overline{S}_{t_0 t}(\lambda_e, n_e) = A_e B_{et_0}^{-1} \lambda_e (b_e n_e)^{-1}.$$

But along an efficient path all the available labor has to be used, so that the amount of resource which is required, denoted by $\hat{S}_{t_0}(n_e)$, is equal to:

$$\hat{S}_{t_0}(n_e) \equiv \overline{S}_{t_0 \infty}(1 - n_e, n_e) = A_e B_{et_0}^{-1} [1 - n_e](b_e n_e)^{-1}.$$

More precisely what has to be available at time t_0 is what we call a resource potential $R \equiv B_e S$, that is an amount of resource measured in efficiency units given the technological level B_e, equal to:

$$R_{t_0} = B_{et_0} S_{t_0} = A_e [1 - n_e](b_e n_e)^{-1}, \tag{3.11}$$

which does not depend upon the time at which the switch is operated but rather upon the type of stationary state followed, that is n_e. Once the economy is evolving along the balanced path its resource potential remains constant. For $t \geq t_0$ we have:

$$B_{et} S_t = B_{et} e^{b_e n_e (t - t_0)} [\hat{S}_{t_0}(n_e) - \overline{S}_{t_0 t}(1 - n_e, n_e)] = B_{t_0} S_{t_0}.$$

Last it has to be pointed out that the right hand side of (3.11), denoted by $H(n_e)$, is a function strictly decreasing in n_e with:

$$\lim_{n_e \uparrow 1} H(n_e) = 0 \quad \text{and} \quad \lim_{n_e \downarrow 0} H(n_e) = +\infty.$$

hence, for a given R_{t_0}, equation (3.11) has a unique solution, which is itself a decreasing function of R_{t_0} and b_e and an increasing function of A_e.

Proposition 3: In the non-renewable resource economy, at any point of time, the economy may switch to a balanced path defined by n_e solution of (3.11). The higher is the resource potential at the time of the switch, the lower is n_e and the higher is the permanent consumption rate the society can enjoy. Along this path the resource potential is constant.

Optimal Stationary Path

To determine the optimal regular path, let us proceed as follows. First consider (3.1) with $\gamma_{e\lambda,t} = 0$ since λ_e has to be positive. In (3.1) let us substitute for ω_t its value given by (3.2) with $\gamma_{en,t} = 0$ since n_e has to be positive too. We get, for a steady state starting from $t = 0$:

$$A_e^{1-\varepsilon}[1 - n_e]^{-\varepsilon} e^{-\rho t} = \lambda A_e B_{e0}^{-1} e^{-b_e n_e t} + \mu_{et} b_e B_{e0} e^{b_e n_e t}$$

equation which is satisfied together with the equation (3.7) determining the trajectory of μ_{et}, if:

$$n_e = \rho b_e^{-1}, \; \mu_{et} = \mu_{e0} e^{-2\rho t}, \; t \geq 0, \quad \text{and} \quad \mu_{e0} = \lambda \rho^{-1} A_e B_{e0}^{-2}[1 - \rho b_e^{-1}].$$

Then the permanent optimal consumption rate is equal to $c = A[1 - \rho b_e^{-1}]$, whilst the permanent resource potential is equal to $A_e[b_e - \rho](b_e\rho)^{-1}$. Last since we must have $n_e < 1$, then such an optimal state exists if $b_e > \rho$.

Proposition 5: In the non renewable resource economy there exists an optimal steady state iff the productivity in the research sector sufficiently high relative to the social rate of discounts, i.e. iff $b_e > \rho$. If this the case the main characteristics of the economy along this optimal steady state are: $\bar{n}$

$$\bar{\lambda} = 1 - \rho b_e^{-1}, \; \bar{n}_e = \rho b_e^{-1}, \; \bar{c} = A_e[1 - \rho b^{-1}] \text{ and } \bar{R} = A_e[b_e - \rho](b_e\rho)^{-1}. \quad (3.12)$$

It has to be noticed that none of these characteristics is depending upon the elasticity of the marginal utility ε. It comes from the very fact that the optimal balanced path is here a stationary state. In models in which the optimal balanced path is not a stationary path, the instantaneous rate of growth of the characteristics of the economy, which are not constant over time, and the values of those variables, which are constant[7], are all depending upon (see Amigues, Grimaud and Moreaux (2004-a), Amigues and Moreaux (2004-b)).

Evolving along an optimal stationary path is an optimal policy if the initial resource potential has a definite value, i.e. for a very specific subset of the initial values of the state variables B and S (such that their product is equal to $A_e[b_e - \rho](b_e\rho)^{-1}$), that is a subset of values which is of measure 0 in the (B, S) space. Hence the optimal behaviour of the economy starting from any value B_0 and S_0 is not a second order problem. As we shall show in the next subsection, the optimal policy starting from not so special initial values of the state variables is depending upon ε.

Transition Towards the Optimal Steady State

The easiest way to look at the transition towards the optimal stationary state, provided that such an optimal steady state exists, is to examine the problem in the (c, R) plan.

Let us first consider the dynamics of c_t.

Differentiating (3.1) w.r.t. time and assuming that $\lambda_{et} > 0$, that is $\gamma_{e\lambda,t} = 0$, we get:

$$-\varepsilon A_e c_t^{-(1-\varepsilon)} \dot{c}_t e^{\rho t} - \rho A_e c_t^{-\varepsilon} e^{-\rho t} = -\lambda A_e B_{et}^{-2} \dot{B}_{et} + \dot{\omega}_t. \tag{3.13}$$

Differentiating (3.2) w.r.t time and assuming that $n_{et} > 0$ that is $\gamma_{en,t} = 0$, we obtain:

$$\dot{\omega}_{et} = \dot{\mu}_{et} b_e B_{et} + \mu_{et} b_e \dot{B}_{et}. \tag{3.14}$$

In this expression of $\dot{\omega}_t$ let us substitute for $\dot{\mu}_{et}$ its value (3.7) taking into account that $\lambda_{et} = 1 - n_{et}$, and for $\dot{B}_{et}$ its value $b_e n_{et} B_{et}$. Then substituting in (3.13) the expression of $\dot{\omega}_t$ so obtained and for $\dot{B}_{et}$ its value $b_e n_{et} B_{et}$ results in:

$$-\varepsilon A_e c_t^{-(1+\varepsilon)} \dot{c}_t e^{-\rho t} - \rho A_e c_t^{-\varepsilon} e^{-\rho t} = -\lambda A_e B_{et}^{-1}. \tag{3.15}$$

Now note that substituting for ω_{et} in (3.1), its value (3.2) with $\gamma_{en,t} = 0$, we get:

$$A_e c_t^{-\varepsilon} e^{\rho t} = \lambda A_e B_{et}^{-1} + \mu_{et} b_e B_{et}. \tag{3.16}$$

Next substituting for $\mu_{et} B_{et}$ its value λS_t (cf (3.9)), we obtain:

$$\lambda A_e B_{et}^{-1} = A_e c_t^{-\varepsilon} d^{-\rho t} [1 + b_e A_e^{-1} R_t]^{-1} \tag{3.17}$$

Last substituting in the r.h.s of (3.15), for $\lambda A_e B_{et}^{-1}$ its value given by (3.17), we obtain the dynamics of c_t:

$$\dot{c}_t = \varepsilon^{-1} [b_e A_e (A_e + b_e R_t)^{-1} - \rho] c_t. \tag{3.18}$$

Note that the consumption must increase, stay constant or decrease according to the resource potential is lower than, equal to or higher than its steady state level $\bar{R}$, so that the curve $\dot{R}_t = 0$ in the (c, R) plan is the vertical through $(0, \bar{R})$ (see Figure 1).

Figure1
Phase Diagram, vase $b > r$

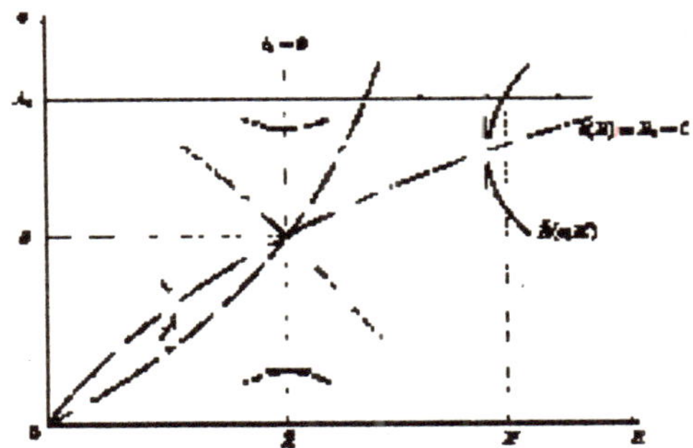

Consider now the dynamics R_t.
Let us differentiate R_t w.r.t time:

$$\dot{R}_t = \dot{B}_{et} S_t + B_{et} \dot{S}_t = b_e n_{et} R_t - B_{et} s_t.$$

But $c_t = B_{et} s_{et}$, so that:

$$\dot{R}_t = b_e n_{et} R_t - c_t$$

and since $n_{et} = [A_e - c_t] A_e^{-1}$, we get the following relation determining the movement of R_t, as far as $n_{et} > 0$:

$$\dot{R}_t = b_e [A_e - c_t] A_e^{-1} R_t - c_t. \tag{3.19}$$

Hence in the (c, R) plan the curve $\dot{R}_t = 0$ is the positive and increasing part of the hyperbola $\hat{c}(R) = b_e A_e R [A_e + b_e R]^{-1}$, through 0, and asymptotic to the horizontal $c = A$ for $R \uparrow +\infty$.

It is easily checked that $\hat{c}(R) = \bar{c}$. It is shown in Appendix A.1 that $(\bar{c}, \bar{R})$ is a saddle-point provided that $b > \rho$.

What remains to be checked is the dynamics of the system starting from a very high resource potential R. One would suspect that for high resource potentials, research today is superfluous and has to be delayed till later.

Let us denote by $\tilde{R}(c; R)$ the function corresponding to the trajectory defined by (3.18)-(3.19), through $\{c = A_e, R > A_e [b_e - \rho](\rho b_e)^{-1}\}$. Then for $c = A_e$ and R satisfying the inequality:

$$d\tilde{R}/dc = -\varepsilon [b_e A_e (A_e + b_e R)^{-1} - \rho]^{-1} > 0 \tag{3.20}$$

and

$$\lim_{R \uparrow \infty} d\tilde{R}/dc = \varepsilon \rho^{-1} > 0. \tag{3.21}$$

By a continuity argument we conclude from (3.20)-(3.21) that the upper part of the stable branch is crossing the horizontal $c = A_e$ at some finite value of and that all the trajectories which first, satisfy the f.o.c's, and second, cross the curve $\tilde{c}(R)$, $R > \bar{R}$, are crossing the horizontal $c = A_e$ too. Thus for $R > \hat{R}$ the optimal policy is to abstain from any research aiming at improving the resource potential, or more modestly to slow down its decrease. The best is to consume, just to consume.

Optimal Policy in the Case of an Insufficiently High Productivity in the Research Sector

It remains to examine what must happen in economies in which the factor productivity in the research sector is low, we mean lower than the impatience rate or social discount rate, ρ. The very fact that the instantaneous marginal utility is going up to infinity for consumption rates decreasing down to zero, implies that, absent any research, in the pure cake eating problem the society would then have to solve, the use of the resource would decrease asymptotically down to zero. Under the strict factor complementarity in the consumption good production industry, the employment in this sector would go down to zero. The best would then be to employ this larger and larger fraction of iddle labor, in the research sector, in order to slow down the speed at which the consumption rate is decreasing. Thus what is happening in case of a rate of impatience larger than the productivity in the research sector is that the consumption goes down to zero although more and more labor is allocated to research. Asymptotically almost all the available labor is employed in the research sector to maintain a very low, and lower and lower, consumption rate. The crux of this dynamics is that at each point of time, a relatively high consumption rate induced by the relative high rate of impatience, depletes the resource stock at a higher proportional rate than the proportional rate at which the resource productivity is increasing.

OPTIMAL DEVELOPMENT IN THE RENEWABLE ECONOMY

We first determine the main characteristics of the optimal plans and next we show that the least research effort policy is the best one.

Main Characteristics of the Optimal Plans

Let LR_t be the Lagrangian of the program (PR):

$$LR_t = (1-\varepsilon)\,(A_r\lambda_{rt})^{-(1+\varepsilon)}\,e^{-\rho t} + v_t[\bar{r} - A_r B_{rt}^{-1}\lambda_{rt}] + \mu_{rt} b_r n_{rt} B_{rt} + \omega_{rt}[1-\lambda_{rt}-n_{rt}] + \gamma_{r\lambda,t}\lambda_{r,t} + \gamma_{rn,t}n_{er}.$$

The F.O.C's are:

$$\partial LR_t/\partial\lambda_{rt}=0 \Leftrightarrow A_e u'(c_t)e^{-\rho t} = A_r c_t^{-\varepsilon}e^{-\rho t} = A_r^{1-\varepsilon}\lambda_{rt}^{-\varepsilon}e^{-\rho t} = v_t A_r B_{rt}^{-1} + \omega_{rt} - \gamma_{r\lambda,t} \quad (4.1)$$

$$\partial LR_t/\partial\lambda_{rt}=0 \Leftrightarrow \mu_{rt} b_r B_{rt} = \omega_{rt} - \gamma_{rn,t} \quad (4.2)$$

together with the complementarity slackness conditions:

$$v_t \geq 0 \quad \text{and} \quad v_t[\bar{r} - A_r B_{rt}^{-1}\lambda_{rt}] = 0 \quad (4.3)$$

$$\omega_{rt} \geq \quad \text{and} \quad \omega_{rt}[1-\lambda_{rt}-n_{rt}] = 0 \quad (4.4)$$

$$\gamma_{r\lambda,t} \geq 0 \quad \text{and} \quad \gamma_{r\lambda,t}\lambda_{rt}] = 0 \quad (4.5)$$

$$\gamma_{rn,t} \geq 0 \quad \text{and} \quad \gamma_{rn,t}n_{rt}] = 0. \quad (4.6)$$

The dynamics of the costate variable must satisfy:

$$\dot{\mu}_{rt} = \partial LR_t/\partial B_{rt} \Leftrightarrow \dot{\mu}_{rt} = -v_t A_r B_{rt}^{-2}\lambda_{rt} - \mu_{rt} b_r n_{rt}. \quad (4.7)$$

Last the transversality condition at infinity is:

$$\lim_{t\uparrow\infty} \mu_{rt} B_{rt} = 0. \quad (4.8)$$

Like in the case of a non-renewable resource, this last condition implies that the imputed value of the stock of technical knowledge has to go down to 0 at infinity.

Let us show now that here the imputed value of the stock of technical knowledge B_{rt} has to be equal to the imputed capital value of the flow of the resource $\bar{r}$.

Proposition 6: In the renewable resource economy, along an optimal path, at each point of time the imputed capital value of the available flow of the resource and the imputed value of the technical knowledge stock must be the same:

$$\bar{r}\int_t^{\infty} v_\tau d\tau = \mu_{rt} B_{rt}, \quad t \geq 0. \tag{4.9}$$

Proof: Multiplying (4.7) by B_{rt} we get:

$$\dot{\mu}_{rt} B_{rt} = v_t A_r B_{rt}^{-1} \lambda_{rt} - \mu_{rt} b_r n_{rt} B_{rt}$$

and since $\dot{B}_{rt} = b_r n_{rt} B_{rt}$, then:

$$\dot{\mu}_{rt} B_{rt} + \mu_{rt} \dot{B}_{rt} = (\mu_{rt}{}^{\bullet} B_{rt}) = -v_t A_r B_{rt}^{-1} \lambda_{rt},$$

hence, since $v_t \bar{r} = v_t A_r B_{rt}^{-1} \lambda_{rt}$ by (4.3), we obtain:

$$(\mu_{rt}{}^{\bullet} B_{rt}) = -v_t \bar{r}. \tag{4.10}$$

Integrating over $[t, \infty]$ we get:

$$\int_t^{\infty} (\mu_{rt}{}^{\bullet} B_{rt}) d\tau = \mu_{rt} B_{rt} - \lim_{t\uparrow\infty} \mu_{rt} B_{rt} = \int_t^{\infty} v_\tau d\tau,$$

By the transversality condition $\lim_{t\uparrow\infty} \mu_{rt} B_{rt} = 0$, hence (4.9).

Now let us multiplying (4.1) by λ_{rt}. Taking into account that $\gamma_{r\lambda,t}\lambda_{rt} = 0$ according to (4.5), that $A_r \lambda_{rt} = c_t$ and that $v_\tau A_r B_{rt}^{-1} \lambda_{rt} = v_t \bar{r}$ by (4.3), we get:

$$c_t(c_t) e^{-\rho t} = v_t \bar{r} + \omega_{rt} \lambda_{rt},$$

hence:

$$MW_t(C) = \bar{r}\int_t^{\infty} v_\tau d\tau + \int_t^{\infty} \omega_{rt} \lambda_{rt} d\tau$$

or, given (4.9):

$$MW_t(C) = \mu_{rt} B_{rt} + \int_t^{\infty} \omega_{rt} \lambda_{rt} d\tau.$$

Proposition 7: In the renewable resource economy there exist two way to decompose the value, in terms of discounted marginal utility, of any optimal consumption plan C:

$$MW_t(C) = \bar{r}\int_t^{\infty} v_\tau d\tau + \int_t^{\infty} \omega_{rt} \lambda_{rt} d\tau = \mu_{rt} B_{rt} + \int_t^{\infty} \omega_{rt} \lambda_{rt} d\tau.$$

Optimality of the Least Research Effort Policy

We define the least research effort policy as the policy consisting to allocate to the research sector this part of the available labor which cannot be employed in the consumption good production industry so long as $B_{rt} < \bar{B}_r$, that is:

$$\lambda_{rt} = A_r^{-1} B_{rt} \bar{r} \quad \text{and} \quad n_{rt} = 1 - A_r^{-1} B_{rt} \bar{r} \tag{4.11}$$

hence:

$$\dot{B}_{rt} = b_r [1 - A_r^{-1} B_{rt} \bar{r}] B_{rt}. \tag{4.12}$$

The solution of this well-known differential equation is the following logistics:

$$B_{rt} = \{\bar{r} A_r^{-1} + (B_{r0}^{-1} - A_r^{-1}\bar{r}) e^{-b_r t}\} = \{\bar{B}_r^{-1} + (B_{r0}^{-1} - \bar{B}_r^{-1}) e^{-b_r t}\}^{-1}.$$

Thus under this policy the critical level $\bar{B}_r$ is attained only asymptotically at infinity so that the resource constraint is always tight. Let us show that this is the optimal policy.

Proposition 8: In the renewable resource economy, whatever the relative values of the research productivity index b and the discount factor ρ, the optimal research policy is the least effort policy.

Proof: We show that, given that the policy is the least effort policy, there exist values of the costate variables and multipliers such that all the optimality conditions (4.1) to (4.8) are satisfied. Before entering the hard core of the proof, it is useful to have in mind the following properties of this policy.

Preliminary Remarks

Let us assume that the policy is the least effort policy, then:

(i) Substituting for $A_r^{-1}\bar{r}$ its value $\bar{B}_r^{-1}$ in the expression of λ_{rt}, we may express λ_{rt} as:

$$\lambda_{rt} = \bar{B}_r^{-1} B_{rt}, \quad t \geq 0, \tag{4.14}$$

(ii) Multiplying the r.h.s of (4.13) by $\bar{B}_r$, we get the following expression of B_{rt}:

$$B_{rt} = \bar{B}_r [1 + (\bar{B}_r B_{rt}^{-1} - 1) e^{-b_r t}]^{-1}, \quad t \geq 0, \tag{4.15}$$

hence:

$$B_{rt} = \bar{B}_r [1 + (\bar{B}_r B_{rt}^{-1} - 1) e^{-b_r(\tau - t)}]^{-1}, \quad \tau > t \quad \text{and} \quad t \geq 0, \tag{4.16}$$

so that:

$$\partial B_{rt} / \partial B_{rt} = B_{rt}^2 B_{rt}^{-2} e^{-b_r(\tau - t)}, \quad \tau > t \quad \text{and} \quad t \geq 0 \tag{4.17}$$

Furthermore:

$$\frac{\partial B_{rt}}{\partial t} - \frac{\bar{B}_r [-\bar{B}_r B_{rt}^{-2} \dot{B}_{rt} + b_e (\bar{B}_r B_{rt}^{-1} - 1) e^{-b_r(\tau - t)}}{[1 + (\bar{B}_r B_{rt}^{-1} - 1) e^{-b_r(\tau - t)}]^2}.$$

In this expression let us substitute for $\dot{B}_{rt}$ its value $b_r n_{rt} B_{rt}$. We get:

$$\frac{\partial B_{rt}}{\partial t} - \frac{b_r \bar{B}_r [\bar{B}_r B_{rt}^{-1}(1-n_{rt}) - 1] e^{-b_r(\tau-t)}}{[1+(\bar{B}_r B_{rt}^{-1} - 1)e^{-b_r(\tau-t)}]^2}.$$

Last substituting for $1 - n_{rt} = \lambda_{rt}$, its value given by (4.13), results in:

$$\partial B_{rt}/\partial t = 0, \ \tau \geq t \quad \text{and} \quad t \geq 0. \tag{4.18}$$

Values of the multipliers for which (4.1) to (4.8) are satisfied
Since here $c_t = B_{rt}\bar{r}$, then V_t (cf (2.8)) is equal to:

$$V_t = (1-\varepsilon)^{-1} \int_t^\infty (B_{rt}\bar{r})^{1-\varepsilon} e^{-\rho(\tau-t)} d\tau \tag{4.19}$$

where B_{rt} is given by (4.15).

If the least effort policy is the optimal policy, then:

$$\mu_{rt} = e^{-\rho t} \partial V_t / \partial B_{rt} \tag{4.20}$$

that is:

$$\mu_{rt} = \bar{r}^{1-\varepsilon} e^{b_r t} B_{rt}^{-2} \int_t^\infty B_{r\tau}^{2-\varepsilon} e^{-(b_r+\rho)\tau} d\tau > 0, \ t \geq 0. \tag{4.21}$$

Note that , implies that , hence:

$$B_{rt}^{-2} \int_t^\infty B_{r\tau}^{2-\varepsilon} e^{-(b_r+\rho)\tau} d\tau \leq (\bar{B}_r B_{r0}^{-1})^2 B_{r0}^{-\varepsilon} \int_t^\infty e^{-(b_r+\rho)\tau} d\tau = (\bar{B}_r B_{r0}^{-1})^2 B_{r0}^{\varepsilon}(b_r + \rho)^{-1} e^{-(b_r+\rho)t}$$

so that:

$$\lim_{t\uparrow\infty} \mu_{rt} \leq \bar{r}^{(1-\varepsilon)}(\bar{B}_r B_{r0}^{-1})^2 B_{r0}^{-\varepsilon}(b_r + \rho)^{-1} \lim_{t\uparrow\infty} e^{-\rho t} = 0. \tag{4.22}$$

Thus the imputed value (4.19) of the stock of technical knowledge decreases to 0 at infinity.

The least effort policy is a policy for which $n_{rt} > 0$ hence $\gamma_{rn,t} = 0$, so that, by (4.2):

$$\omega_{rt} = b_r \mu_{rt} B_{rt} > 0. \tag{4.23}$$

Let us now determine the value of v_t for which, if μ_{rt} is given by (4.20), then (4.3) and (4.7) are satisfied.

From (4.21), we get:

$$\mu_{rt} B_{rt} = \bar{r}^{1-\varepsilon} e^{b_r t} B_{rt}^{-1} \int_t^\infty B_{r\tau}^{2-\varepsilon} e^{-(b_r+\rho)\tau} d\tau.$$

Let us define J_t as follows:

$$J_t = e^{b_r t} B_{rt}^{-1} \int_t^\infty B_{r\tau}^{2-\varepsilon} e^{-(b_r+\rho)\tau} d\tau \Rightarrow \bar{r}^{1-\varepsilon} = J_t. \tag{4.24}$$

Then, given (4.10), we must have:

$$v_t = -\bar{r}^{-\varepsilon} \dot{J}_t. \tag{4.25}$$

What we have to show is that $v_t \geq 0$, that is $\dot{J}_t \leq 0$ [8].

Differentiating w.r.t time, taking into account that $\partial B_{rt}/\partial t = 0$ and that $\dot{B}_{rt} = b_r n_{rt} B_{rt}$, we get:

$$\dot{J}_t = (1 - n_{rt}) b_r e^{b_r t} B_{rt}^{-1} \int_t^{\infty} B_{r\tau}^{2-\varepsilon} e^{-(b_r+\rho)\tau} d\tau - B_{r\tau}^{1-\varepsilon} e^{-\rho t}. \qquad (4.26)$$

Let us consider successively the cases $\varepsilon \geq 2$ and $\varepsilon < 2$.

If first $\varepsilon < 2$, then:

$$\tau \geq t \Rightarrow B_{rt}^{2-\varepsilon} \leq B_{r\tau}^{2-\varepsilon} < \overline{B}_r^{2-\varepsilon},$$

hence:

$$\dot{J}_t \leq b_r e^{b_r t} \overline{B}_r^{-1} \int_t^{\infty} \overline{B}_r^{2-\varepsilon} e^{-(b_r+\rho)\tau} d\tau - B_{rt}^{1-\varepsilon} e^{-\rho t},$$

so that:

$$\dot{J}_t \leq e^{-\rho t} \overline{B}_r^{1-\varepsilon} [b_r (b_r + \rho)^{-1} - (B_{rt} \overline{B}_r^{-1})^{1-\varepsilon}].$$

Last $\varepsilon \geq 2$ and $B_{rt} < \text{B}_r$ imply that $(B_{rt} \overline{B}_r^{1})^{1-\varepsilon} > 1$ so that $\dot{J}_t \leq 0$.

If second $\varepsilon \geq 2$, then:

$$\tau \geq t \Rightarrow \overline{B}_r^{2-\varepsilon} < B_{r\tau}^{2-\varepsilon} \leq B_{rt}^{2-\varepsilon},$$

hence:

$$\dot{J}_t \leq b_r e^{b_r t} \overline{B}_r^{-1} B_{rt}^{2-\varepsilon} \int_t^{\infty} e^{-(b_r+\rho)\tau} d\tau - B_{rt}^{1-\varepsilon} e^{-\rho t} = e^{-\rho t} B_r^{1-\varepsilon} [b_r (b_r + \rho)^{-1} - B_{rt} \overline{B}_r^{-1} - 1] < 0.$$

What remains to show is that (4.1) is satisfied for ω_{rt} given by (4.23) and v_t given by (4.25).

In (4.23) let us substitute for μ_{rt} its expression (4.21). We get:

$$\omega_{rt} = b_r \bar{r}^{1-\varepsilon} e^{b_r t} I_t B_{rt}^{-1}, \qquad (4.27)$$

where:

$$I_t = \int_t^{\infty} B_{\tau}^{2-\varepsilon} e^{-(b_r+\rho)\tau} d\tau. \qquad (4.28)$$

From (4.25) and (4.26), taking into account that $1 - n_{rt} = I_{rt}$, we obtain:

$$v_t = -\bar{r}^{-\varepsilon} [\lambda_{rt} b_r e^{b_r t} B_{rt}^{-1} I_t - B_{rt}^{-1} e^{-\rho t}]. \qquad (4.29)$$

Last, since $c_t = B_{rt} \bar{r}$, then:

$$A_r c_t^{-\varepsilon} e^{-\rho t} = \bar{r}^{-\varepsilon} A_r B_{rt}^{-\varepsilon} e^{-\rho t}. \qquad (4.30)$$

In (4.1) let us substitute for ω_{rt}, v_t and $A_r c_t^{-\varepsilon} e^{-\rho t}$ their expressions (4.27), (4.29) and (4.30) respectively, to obtain:

$$\bar{r}^{-\varepsilon} A_r B_{rt}^{-\varepsilon} e^{-\rho t} + \bar{r}^{-\varepsilon} [\lambda_{rt} b_r e^{b_r t} B_{rt}^{-1} I_t - B_{r\tau}^{-1} e^{-\rho t}] A_r B_{rt}^{-1} - b_r \bar{r}^{1-\varepsilon} e^{b_r t} B_{rt}^{-1} I_t$$

$$= \bar{n}^{-\varepsilon} \lambda_{rt} b_r e^{b_r t} I_t A_r B_{rt}^{-2} - b_r \bar{r}^{1-\varepsilon} e^{b_r t} B_{rt}^{-1} I_t$$

Since $A_r \lambda_{rt} = B_{rt} \bar{r}$, this last expression may be rewritten as:

$$b_r \bar{r}^{1-\varepsilon} e^{b_r t} B_{rt}^{-1} I_t - b_r \bar{r}^{1-\varepsilon} e^{b_r t} B_{rt}^{-1} I_t = 0$$

so that (4.1) is satisfied.

CONCLUSION

We have shown that, starting from initial conditions corresponding to situations in which the resources are rare, the optimal research policies, in the non renewable and renewable resource economics, are strongly contrasted, as far as the resource dedicated research is concerned. In the non renewable resource economy the type of policy is depending upon the relative values of the productivity index in the research sector and of the discount rate, whereas in the renewable resource economy, whatever the relative values of b and ρ, the type of research policy is always the same. Assuming that the research effort productivity is sufficiently high, that is $b > \rho$, in the stationary state of the non-renewable resource economy a permanent effort has to be maintained in order to balance the resource stock decrease. On the contrary in the renewable resource economy no research effort has to be done in the stationary state[9]. This robust conclusion, which does not depend upon the shape of the production function in the consumption good industry. The point, which has to be forcefully emphasized here, is the crucial difference between increases of the resource productivity in the two kinds of economies. In the non-renewable resource economy such an increase is equivalent to a finite increase of the resource endowment, the intertemporal use of which is free and has to be consumed optimally. In the renewable resource economy a marginal increase in the resource productivity corresponds to an infinite increase in the resource endowment, the use of which is constrained to a marginal increase of endowments available at each point of time, so that to these infinite increase there corresponds a finite increase in welfare, an immediate byproduct of the discounting of the instantaneous utilities. The next step in our research agenda is to compare these two types of economies for less extreme production technologies.

APPENDIX

In order to show that $(\bar{c}, \bar{R})$ is a saddle point, let us consider the system of differential equations determining the trajectories of c_t and R_t:

$$\dot{c}_t = \varepsilon^{-1}[b_e A_e (A_e + b_e R_t)^{-1} - \rho]\, c_t, \tag{6.1}$$

$$\dot{R}_t = b_e [A_e - e_t] A_e^{-1} R_t - c_t. \tag{6.2}$$

The derivatives of $\dot{c}_t$ and $\dot{R}_t$ are:

$$\partial \dot{c}_t / \partial c_t = \varepsilon^{-1}[b_e A_e (A_e + b_e R_t)^{-1} - \rho], \;\; \partial \dot{c}_t / \partial R_t = -\varepsilon^{-1}(b_e^2 A_e c_t)(A_e + b_e R_t)^{-2}$$

$$\partial \dot{R}_t / \partial c_t = -b_e A_e^{-1} R_t - 1, \;\; \partial \dot{R}_t / \partial R_t = b_e [A_e - c_t] A_e^{-1}.$$

The Jacobian of the system at the stationary point $(\bar{c}, \bar{R})$ is:

$$J_a = -\rho[b_e - \rho]\,\varepsilon^{-1},$$

which is negative provided that there exists a stationary path, that is $b_e > \rho$.

The characteristic equation is:

$$P(x) = x^2 = \rho x - \rho[b_e - \rho]\varepsilon^{-1} = 0,$$

the roots of which are real and of opposite signs if $b_e > \rho$. Hence $(\bar{c}, \bar{R})$ is a saddle point.

NOTES

1. INRA and University of Toulouse 1 (IDEI and INRA-LERNA), 21 Allée de Brienne - 31000 Toulouse, France. *E-mail: amigues@toulouse.inra.fr*
2. Université de Toulouse I, (IUF, IDEI and INRA-LERNA), 21 allée de Brienne- 31000 Toulouse, France. *E-mail: mmoreaux@cict.fr*
3. cf Wrigley (1988).
4. cf Wrigley, *ibid*. Admittedly Wrigley does not evoquate the nuclear society
5. See Amigues *et alii* (2004).
6. As usual in this type of program we investigate only the set of feasible programs for which S_t ¤ 0. Hence this constraint is deleted in the Lagrangian.
7. Like for example the sectoral employment if the rate of growth of the labour supply is equal to 0.
8. Under the least effort policy $\bar{r} = A_r B_{rt}^{-1} \lambda_{rt} = 0$ by definition, so that what has to be is that ✠$_t$ ¤ 0 in order that (4.3) be satisfied.
9. For other differences in the two types of economies see also Moreaux and Ricci (2000, 2004).

REFERENCES

Amigues, P. P., N. V. Long and M. Moreaux (2004), Overcoming the Natural Resource Constraints Through Dedicated R&D Efforts: Contrasting the Non-Renewable and the Renewable Resource Economies, forthcoming in *International Journal of Global Environmental Issues*.

Amigues, J-P. Grimaud A. and M. Moreaux (2004-a), Optimal Endogenous Sustainability with an Exhaustible Resource Through Dedicated R&D, Mimeo, LERNA-Université de Toulouse 1.

Amigues, P-P. and M. Moreaux (2004-b), Endogenous Growth with an Exhaustible Resource: A Bisectoral R&D Model, Mimeo, LERNA-Université de Toulouse 1.

Moreaux, M. and F. Ricci (2000), Developing Resources from Resources, LERNA D. P. 001.11.44, Université de Toulouse 1.

Moreaux, M. and F. Ricci (2004), The Simple Analytics of Developing Resources from Resources, forthcoming in *Resources and Energy Economics*.

Wrigley, F. A. (1988), *Continuity, Chance and Change – The Character of the Industrial Revolution in England*, Cambridge: Cambridge University Press.

22

COMMON PROPERTY RESOURCES MANAGEMENT IN INDIA: SOME REVIEW FINDINGS

Naresh Chandra Sahu and Binayak Rath

INTRODUCTION

Common property is that which is held and used jointly by a group or community. Broadly, common property resources (CPRs) are those resources, which are accessible to the whole community to which no individual has exclusive property rights (Jodha, 1986). Apart from these, the use of these resources by the users are subtractive and having the characteristic of a public good such as indivisibility. Adhering to the definition of Jodha, the resources falling under this category of CPRs in India are protected forests, community forests, village pastures, waste lands, waste dumping places, watershed drainages, grazing land (gauchar), village ponds, canals, tanks, wells, streams, rivulets, and river beds. However, CPRs are regarded as an important resource base for the rural economy in many of the developing and underdeveloped countries till today because of their significant economic contributions to the sustenance of rural livelihood (Jodha, 1986). CPRs render substantial contribution to the rural livelihoods, especially to the small as well marginal farmers and poor labourers. The contributions include physical supply of products like food, fuel, fodder, timber, medicinal plants, manure, silt, and water. In addition, the other contributions of CPRs are: widening the range of income –generating activities available to the people in rural areas and to provide inputs to households for domestic consumption and production; supplying inputs to the agro based cottage and handicrafts industries; improving physical environment of flora and fauna; and safety net for the poor in terms of income generation and food supply.

The common property of resources in developing countries have increasingly been facing lots of problems of degradation, decline and destruction of over the last few decades due to the alarming growth of population, commercialization, industrialization and urbanization. However, several scholars have identified four important factors responsible for the deterioration of the status of the CPRs in India are: (1) encroachment due to population growth; (2) legal privatization (institutional change in the form of land reform policy in the early 1960s); (3) poor management, (4) industrialization and infrastructure development and (5) commercialization of the CPRs based activities.

The outcome of the destruction and degradation of CPRs have adverse impact not only on the ecological stability but also livelihood pattern of the rural people who depend heavily on it for meeting their numerous necessary requirement such as food, fuel, fruits, fodder, medicinal plants, housing building materials, water, and a host of other things daily. Moreover, the environmental stressed migration, drying tanks, irregular monsoon, and soil erosion are glaring instances of the decline and degradation of the CPRs in India. Similarly, the livestock farming which the key activity is sustained by CPRs are largely affected due to the decline of these resources. However, women fuel gatherers, particularly from tribal belts and lower strata of the society, have to walk miles in a day to collect a head load of fuel wood to boil the pot of their family due to the massive deforestation. In order to reduce the degradation and depletion forest resources government of India drastically changed its policy from state property regime to community based forest resource management in late 1980s. The failure of state property regime promoted government to declare a new and people oriented national forest policy in 1988, in which the participation of the people was encouraged and ensured property rights them over forests. This is very much known as the Joint forest management (JFM) system and under this system many forest protection committees are formed to protect and develop the degraded forests. Till today there are around 39270 JFM committees covering 1048383 hectares of forest area of the country.

Recently, some scholars have come up with many interesting findings from various parts of the country pertaining to the impacts of JFM committees on status of local environment, livelihood pattern of the users, and the management structures in details. Even some scholars strongly posit the view that efficient management of CPRs is very important for poverty alleviation, reduction of distress migration, sustainable development with equity, ecological stability and preserving the biodiversity (Naik, 1997; Saxena, 2000; Heltberg, 2001). Therefore, the best approach advocated many scholars is to adopt participatory approach and build community based resource management institutions for better and sustainable management of the CPRs, such as forests by ensuring sufficient incentives and mentoring mechanisms to make these institutions long lasting and sustainable in the years to come.

Besides, there are some studies, which put forward some confusing results while empirically testing the impacts of CPR institutions in terms of social equity and welfare as a whole. However, an attempt has been made in this paper to critically review the various studies of scholars pertaining to the CPRs management institutions such as JFM committees[1], Forest Protection Committees (FPCS), etc and important issues associated with them in different parts of India under the aegis of Joint Forest Management regime over the last couple of decades.

COMMON PROPERTY RESOURCE DEBATE

Recent literature on CPR management not only criticized "Hardin's Tragedy of the Commons" (1968) but also has established that the tragedy of commons often results from institutional failure to control access to resources rather than the inherent failure of common property. It argues that tragedy of commons can be ruled out provided the society is capable of enforcing the collective use norms.

Some critics are of the opinion that Hardin's Tragedy of Commons is applicable only to the situation of appropriation of 'open access resources' but not to common property resources where property rights are well defined and well implemented (Ciriacy-Wantrup and Bishop, 1975). In case of open access system the individuals do not get proper incentives to act in a socially and economically efficient manner. In the literature of common property resources, broadly three different schools of thought have emerged on institutional arrangements to avert the tragedy of commons as envisaged by Hardin. According to property rights school, the problem of overexploitation and degradation of common property resources (CPRs) can be resolved only by creating and enforcing private property rights. Private property is considered to be the most efficient way to internalize the externalities generated from overexploitation of the commons. Scholars of second school of thought advocate that state property regime can reduce the overexploitation and degradation of CPRs (Hardin, 1968). Institution building at the local community level for managing the common pool resources has emerged as a third possibility in due course of time. An increasing number of scholars, however, advocate that decentralized collective management of CPRs by their users and stakeholders could be one of the best alternative systems for overrating the 'tragedy of commons' (Wade, 1987; Jodha, 1986, 1995; Chopra *et al.*, 1989; Ostrom, 1990). They argue that privatization of CPRs would have adverse impacts on the local population who are usually dependent on them to meet their basic requirements. It will further enhance the social inequity. Therefore, privatization of CPRs cannot be a sustainable means to manage those who have positive externalities from social welfare perspectives.

Supporting the arguments of the scholars who believe in the strength of institutions for the management of CPRs, Ostrom (1990) suggested that the collective action for CPRs management will be long enduring and successful under conditions of well defined boundaries, harmony between appropriation and provision rules, proper sanctions, efficient conflict-resolution mechanisms, proper incentives and effective monitoring. In other words, the proposition suggested by her exclusively calls for a well defined and sanctioned community based property rights regime resulting from the formation of community based institutions for the sustainable management of the commons.

Moreover, it is argued property rights by themselves do not provide adequate incentives and conditions for sustainable management of CPRs. Appropriate cost-benefit sharing arrangements along with empowerment of the users or stakeholders, technical assistance to develop and strengthen local organizational capacities will promote equitable and sustainable management efforts. The very success of the property right regime depends upon harmony of ecosystem and governance boundaries, specification and representation of interests, matching of governance structures to ecosystem characteristics, containment of transaction costs, and the establishment of monitoring, enforcement and adoption processes at the appropriate scale (Ostrom, 1990; Bromely, 1991). While the aggregate gains from reducing common pool problems or promoting economic growth through the definition or redefinition of property rights are unlikely to be controversial, the distribution of wealth and political power inherent in the proposed rights structure will be a source of disputes. Restricting the access of the poor people to natural resources through changes in property rights structure in common pool resources is likely to increase the level of poverty unless specific measures of compensatory transfer schemes are undertaken to safeguard the interests of the most of the most vulnerable section of the community (Adhikari, 2002). There seems no reason to suppose *a priory* that institutions are always efficient and equitable and they serve the purpose for which the institutions were created. Therefore, the issues and various aspects related to the community based resource management institutions should thoroughly be studied for shaping the policies and planning relating to the sustainable management of the natural resources.

PARTICIPATORY MANAGEMENT OF FOREST RESOURCES

The changes in the management structure of CPRs, such as forests over time in India have been manifested in the different forest policies, which outline the access rights of various interest groups, users and stakeholders. The post independence national forest policy 1952 reiterated the principles of colonial forestry, as enshrined in the earlier forest policy resolution of 1894, and reinforced the exclusive rights of the state with regard to the management and control of forest resources. A goal was adopted of bringing 60 per cent of the area in hilly tracts and 20 per cent of the area in plains (one-third of the country's total geographical area) under forest cover for ecological reasons. The commercial exploitation of the forests was given thrust by declaring as vital national need. Further, the relevance of the forests to meet the needs of defence, reconstruction schemes such as river valley projects, development of industries and communication was asserted in this forest policy.

However, it has been realized that the state control or say, state property rights (absence of local user's participation) could not prevent the forests resources from higher degradation and depletion of natural resources against the backdrop of increasingly pressure from the users. The degradation and depletion of forest was very rampant during the period from 1952 – 1988 because the local users were sidelined in top down approach policy of the government to manage the forests, which indirectly encouraged them to exploit the forests resources as the open access resources. The state property regime in the management of forests could not prevent massive degradation and manage forest resources sustainably due to lack of scope of local users participation in the management of forests

The failure of the top down approach adopted by the Government induced them to search for alternatives. At the same time the success of forest development projects with participation of people, viz. Chipko[2] Movement, made government think about the significance of the people's participation in the management arrangements of forest resources. In view of some success stories from different parts of the country Government of India adopted the participatory management of forest resources since 1988 and accordingly ensured property rights to community who are living nearby the forested areas.

In India like most of the developing countries, the community based resource management institutions, such as Joint Forest Management (JFM) committees, have been encouraged since late 1980s and early 1990s in order to achieve the better management CPRs. Though the ownership rights and control rested with the state but the policy at least recognized the rights of stakeholders and multiple users to meet the daily requirements such as fodder, fuel, and non-timber forest products without depleting and degrading the status of forest. The basic objectives of the National Forest Policy, 1988 are:

- Maintenance of environmental stability through preservation and, where necessary, restoration of the ecological balance that has been adversely disturbed by serious depletion of the forests of the country.
- Conserving the natural heritage of the country by preserving the remaining natural forests with the vast variety of flora and fauna,

which represents the remarkable biological diversity and genetic resources of the country.

- Checking soil erosion and denudation in the catchment areas of rivers, lakes and reservoirs in the interest of soil and water conservation, for mitigating floods and droughts and for the retardation of siltation of reservoirs. Also checking the extension of sand dunes.
- Increasing substantially the forest/tree cover in the country through massive afforestation and social forestry programmes.
- Meeting the requirements of fuel wood, fodder, minor forest produce and small timber of the rural and tribal population.
- Increasing the productivity of forests to meet essential national needs.
- Encouraging efficient utilisation of forest produce and maximizing substitution of the wood.
- Creating a massive people's movement with the involvement of women, for achieving these objectives and to minimize pressure on existing forests.

However, the ground for people's participations in the management of forest was created by recognizing the rights of stakeholder s over forests in this policy. Therefore this policy is regarded as a landmark in the beginning of people's participation in the protection, management, and development of forests in India.

The goal is to promote stakeholders' active involvement in management of protected areas and other natural resources. The major shift in the paradigm of natural resources management from state to community based resources management has empowered the local users to manage CPRs in efficient way. Similarly, the government of India has recognized the community-based management of natural resources in its forest policy of 1988. Since then the Joint Forest Management (JFM) has been initiated with great vigour in India for involving local people (stakeholders) in forest management and envisaged a formidable partnership between the people and state government to protect and regenerate forest while meeting people's needs in a sustainable manner. This shift in policy is no more than a belated recognition of that sustainability of collective human institutions to move towards sustainability. In such resource governance framework the local users should have the greatest stakes in the management of these resource institutions (Agrawal, 2001). In parallel, the recent literature on CPRs management emphasizes the ability of user communities to effectively manage collective owned natural resources through informal institutional arrangements (e.g. Wade, 1987; Ostrom, 1990).

PROGRESS OF PARTICIPATORY FOREST MANAGEMENT IN INDIA

Owing to National Forest Policy 1988, the Government of India issued a notification in June 1990[3] requesting the State Governments to involve local communities in the management of forests. It is envisaged that the communities, in lieu of their participation in protection and development of forest areas, will be entitled to sharing of usufructs in a manner specified by the concerned State Forest Departments. This has led to the development of Joint Forest Management (JFM) programme. As result, the state governments of the country issued their own resolution while enabling the respective forest department to encourage the people to form JFM committees to regenerate the degraded forests in a sustainable manner.

Presently, there are 22 State Governments have issued resolutions in this regard. However, the state governments have evolved their own mechanisms of involving local communities in conformity with the proclaimed policy. The local institutions engaged in this programme are known by different names in different states like Forest Protection Committee (FPC), Village Forest Committee (VFC), *Van Samrakshan Samiti*

Table 1
Area under Joint Forest Management

State	*Date of notification*	*No. of JFM Committees*	*Area under JFM ('000ha)*
Andhra Pradesh	28.09.92	6575	1,632.19
Arunachal Pradesh	03.10.97	10	5.29
Assam	10.11.98	101	3.06
Bihar	08.11.90	1675	935.08
Gujarat	13.03.91	706	91.07
Haryana	13.06.90	350	60.73
Himachal Pradesh	12.05.93	203	62.00
Jammu & Kashmir	19.03.92	1599	79.27
Karnataka	12.04.93	1212	12.80
Kerala	16.01.98	21	4.00
Madhya Pradesh	10.12.91	12038	5,800.00
Maharashtra	16.03.92	502	94.73
Mizoram	18.09.98	103	5.87
Nagaland	5.03.97	55	0.65
Orissa	3.08.88	3704	419.31
Punjab	14.07.93	89	38.99
Rajasthan	16.03.91	2705	235.63
Sikkim	26.06.98	98	2.19
Tamilnadu	08.08.97	599	224.38
Tripura	20.12.91	157	16.23
Uttar Pradesh	30.08.97	197	34.59
West Bengal	12.07.89	3431	490.58
Total	36,130	10248.64	

Source: State of Forest Report 1999, Forest Survey of India, Dehradun

(VSS), Village Forest Protection Management Committee (VFPMC) etc. The nature of usufruct sharing also varies from state to state. In constitution of committees, representation of women is also ensured. About 36,130 Forest Protection Committees are managing a total of 10.25 million ha of forest area (State of Forest Report, 1999). The details of number of committees and area under their management are given in following table-1.

A close analysis of the above table reveals that Madhya Pradesh alone accounts for more than 50 per cent of the area under JFM. It is followed by Andhra Pradesh, Bihar, West Bengal and Orissa. The extent of involvement of local communities in the north eastern states is quite limited as at present there are 426 committees managing 31,092 ha area only. In order to assess the impact of JFM programme, it is essential to evaluate the forest areas under JFM and socio- economic status of the participating communities periodically. However, these CPR institutions are yielding positive outcomes in the managment, development and protection of common property resources in spite of getting entangled with numerous problems which vary across the states in the country.

ISSUES AND CHALLENGES FACED IN MANAGEMENT OF FORESTS

Although local control of natural resources is now regarded as a win-win solution for the government, local people and the donor institutions, the empirical evidence regarding the impact of common property institutions is rather insufficient and scanty. There is still insufficient solid empirical knowledge about the evolution and functioning of local NRM institutions and how government and donor interventions can play shape that process (Heltberg, 2001). The current issues and challenges relating to community forestry under the participatory approach in India focus on the following questions:

- What makes some systems of community based management more successful than others in sustaining the resource and distributing its products?
- What production processes are taking place and what contribution do they make to the income of the poor households?
- What are the community and household characteristics that promote inequitable access to and control over resources and local institutions?
- What are the local socio-economic consequences of CPR management institutions?
- How do community based property rights over forests affect the livelihood strategies of the poor?

- Does Transaction cost involved in the management of CPR institutions play any role?
- What is the impact of CPR institutions on the stress migration in the rural area?
- What is the impact of social capital on the CPR institutions?

Though there is extensive literature on success of CPR management systems in conserving local resources, empirical discussions regarding the interplay between poverty, migration, institutional change, and distributional implications of CPR management is not properly reflected A most striking feature of these studies lie in the fact that their authors are generally convinced that, given the glaring failure of state ownership experiences in developing countries, collective, community based regulation holds out the best prospects for an efficient management of village level natural resources. However it is recognized that the balance sheet of actual experience of common property options rather mixed; the central focus on the research should be directed towards understanding the reasons that can account for these varying levels of performance of CPR institutions. It is not clear upon what condition resource management will be optimal to ensure equity and efficiency among the resource users. Further research on determinants and impact of local management institutions and thus performance of CPR management is emphasized mainly in relation to distributional implication of the regime, decision making arrangements and the way individuals interact when applying rules (co-operation, free riding etc.) and an analysis of the outcome (efficiency and equity). So far systematic tests of the relative importance of factors important sustainability, equity, or efficiency of common property institutions are relatively uncommon (Agrawal, 2001). A clear understanding of the local management institutions, their determinants and their role in shaping the environmental outcome is desirable to better guide future NRM policies (Heltbeg, 2001).

DEGRADATION AND DECLINE OF CPRs

In India, as in most of the developing countries, the environmental degradation has manifested itself in rapid rates of natural capital depletion exemplified by the degradation of CPRs such as, forest and soil erosion. There have been serious concerns expressed by numbers of scholars from time to time over the status, size and sustainability of the CPRs since the publication of an essay entitled "Tragedy of Commons" by Hardin in Science Journal in 1968. Broadly, the decline CPRs occur in three ways: physical loss of resources due to the construction of infrastructure, deterioration of physical productivity (i.e., quality) of the resources and changes in the usage and property rights of CPRs. However, in the study of seven states in India as above discussed Jodha (1986) had first identified the principal factors attributed to the decline and degradation of CPRs. The major factors are: population growth, illegal encroachments, market forces, legal privatization, technological changes and environmental stress (e.g., drought). It was found in the study that the legal privatization contributed 40 per cent of decline of the size of common property land resource in the study area. In 1950s and 1960s the Government of India launched several welfare programmes like 'land to poor' in the form of the distribution of common property land through the panchayat to the poor households in the rural area because the allotment of these resources was the easiest way for this purpose. As a result, the allotment of common property land to socially backward classes resulted in the massive decline in the size of the common property land.

In a village level study in Karnataka, Pasha (1992) had also estimated that the area under CPRs declined from 6999 acres (35.6% of the geographical area of the villages covered) to 4654 acres, a decline of 23.7%. The main cause of this decline was privatization by both rich and poor. In the study area, rich households appropriated 52% of the land lost. The poor households obtained 25.6 % of the land lost through Government redistribution programs, and encroached illegally on a further 22.4% of the CPR land. Iyenger (1989) had carried out a very comprehensive village level study to examine the incidence and intensity of the problems faced by the CPRs by covering 25 villages across the different regions in Gujarat. The study, like other studies analyzed above, posited that major reason for the reduction of the size of CPRs was the legal privatization in the study area. Furthermore it was found in the study that the reduction of size and quality of the CPRs had adversely affected the livelihood patterns of the communities in a great way, more particularly the poor households.

Reddy *et al.* (2001) empirically identified the determinants of degradation and decline of CPRs like forest in India adopting ordinary least regression method by using the time series data across the states. The findings of the study show that demographic factors (population density and land man ratio), urbanization (percentage of population living in the urban area and road length), and livestock have positive and significant impact on the forest resource degradation in the states of India. But both the development indicators, namely, relative development index and per capita income have a negative impact on the forest degradation indicating that the forest degradation is not associated with general economic development. In addition, irrigation is also positively associated with the forest degradation and this could be due to the reason that irrigation enhances e productivity of and profitability of the land. Hence the demand for land (pressure on forest) will be more in irrigated regions. Furthermore, the study also suggested that rural literacy and institutional credit have revealed negative impact on forest degradation. These two turned out to be important policy variables for checking the CPRs like forest. Rural literacy not only improves the awareness of the farmers towards environment but also opens up non-farm income avenues, which will reduce their dependence on CPRs.

Most of the studies analyzed above clearly reveal that the poor household's dependence on CPRs is greater in comparison to the rich households to meet their daily livelihood requirements. However, the literature relating environmental degradation in the developing countries is primarily dominated by the *poverty- environment hypothesis* (PEH), which explicitly explains that reduction of poverty is a precondition for environmental sustainability. Recently this hypothesis has received prime attention by scholars and policy experts while discussing the sustainability of CPRs (Jalal, 1993; Barbier, 1997a and b; Lopez, 1998). These studies undertaken by the above mentioned scholars suggest that the poor rely more than others on CPRs owing to variety of reasons: a lower shadow cost of labour (the main source of access cost), a lower preference for cleaner but more expensive fuel substitute, and credit constraints that prevents shift to less forest intensive occupations, livestock and land practices.

However, there have been some literatures on the poverty environment hypothesis (PEH), which assert that the poverty alleviation is a precondition to ease pressure on forests. Baland and Platteau (1996, 1997) had theoretically noted that higher poverty or inequality might inhibit an informal collective action within the rural community that regulates CPR use. According to the PEH, therefore, halting environmental degradation requires the reduction of poverty, via growth or redistributive public policies. The related literature on environmental Kuznets curves based mainly on macro evidence also suggests that increases in the living standards beyond some threshold will ease pressures on environmental resources (Barbier, 1997a and b; Grossman and Krueger, 1995).

On the contrary, Bardhan *et al.* (2002) suggested that the rising standard of living does not reduce pressures (the collection of fuel wood) on the CPRs after studying the household data from the World Bank 1995-95 Living Standard Measurement Survey for 215 Nepal villages by adopting different econometric tests. The study does not provide any evidence for the support of PEH hypothesis, which implies that increasing living standard reduces the pressures on the forests as argued above by several scholars. It was found that the collection of fuel woods could, however, be substantially reduced by spread of primary education and growth of non-farm employment opportunities. Therefore, the PEH hypothesis needs to be carefully examined by academicians before deriving any final conclusion over this issue.

DETERMINANTS OF CPRS MANAGEMENT INSTITUTIONS

With an attempt of reviewing the existing literature on the joint forest management in India, Naik (1997) had identified some of the important factors influencing the household participation in the common property resource institutions such as joint forest management (JFM) committees while undertaking two case studies in Gujarat. In this study the identified determinants of people's participation in JFM are: quality of agricultural land, percentage of irrigated land, size of agricultural landholding, level of technology adoption, quality of forest land, per capita size of forest land,, rainfall, leadership in the village, perception of the forest department policy by the multiple users and stakeholders, share in the forest produce, marketing infrastructure for the forest products, marketing infrastructure for agricultural products, family size, availability of employment opportunities, interest rate, and expected time of flow returns from JFM. In addition, the findings of the study show that poor quality of agricultural, low percentage of agricultural land, higher quality of forest land, larger per capita forest land, erratic rainfall, developed marketing infrastructure for the forestry products, large family size, larger share in the forest produce, low interest rate, and quick expected time of flow of returns from JFM have resulted in the higher participation in the joint forest management institutions in the study area.

Similarly, Heltberg (2001) had analyzed the various factors and dimensions of local institutions for forest conservation and management by using different econometric techniques. The study was based on the primary data collected from thirty villages and hundred eight households which were randomly sampled from a protected area named Sariska Tiger Reserve in Rajasthan. Logit model was used to identify the factors responsible for formation of institutions and the higher probability of collective action in the protection and conservation of forest. The impacts of institutions and other variables on common resource dependence and forest outcomes were tested using Instrumental Regression techniques. The major findings of the study postulate that village population size has positive effect on the probability of collective action whereas the institutional experiences, physical infrastructure measured by development index and resource scarcity as measured by people per unit of resource do not encourage the formation of community based forest resource management institutions. Besides this, the population growth is not a significant determinant of formation of collective action i.e. community based forest resources management in this study. The study also concluded that large landowners are heavily involved in degrading forests especially when these resources have good market value. However, this study has not discussed the distributional implications of the common property institutions on the multiple users and stakeholders in the region studied. In addition, the role of non-governmental organization (NGO) and voluntary agencies in building the CPR institutions are not taken up properly.

CPRs AND EFFICIENCY

The government of India launched social forestry programme in the late 1970s and early 1980s by taking into consideration the recommendations made by National Commission Agricultural (NAC) in order to cater the needs of the people with a view of reducing deforestation and save forest. This programme outlined that the forestry programme would be undertaken in the non-forest land (i.e. private land and village land) for meeting fuelwood and fodder demand in the rural area. The programmes laid down special emphasis on community participation. In fact the prgramme got real boost in the early 1980s when several international funding agencies such as World Bank, the Food and Agriculture Organization of the United Nations (FAO), the United States Agency for International Development (USAID), the Swedish International Development Agency (SIDA), etc started co-financing with the state forest departments social forestry programme in India. The total external assistance for social forestry projects over the decade, 1980-90, was about Rs 10 billion (at Rs 42 = US $1).

However, for some scholars, it was seen as a programme that would release industrial forestry from social pressures. The core objective for forestlands remained the production of the commercial timber production. The programme of social forestry projects has the following four main components: (1) Reforestation and rehabilitation of degraded forestlands; (2) Strip plantation on roadsides, along canal banks and railway lines; (3) Community forestry on village and state- owned land; and (4) Farm forestry. Studies carried out by scholars suggest that the social forestry programme was successful especially in northwest India.

In 1991, Patnaik and Bhartuar, in their article " Ecological Benefits of Five Year Old Social Forestry Village Woodlots in Orissa" had attempted to evaluate the ecological and environmental benefits of five year old plantation in social forestry project villages in Orissa. However, they had discussed the effects of plantations only qualitatively ignoring quantitative assessment. In a study Shukla (1991) had made an attempt to find out the employment and income generation of the of social forestry project in Gujarat state. In Gujarat, social forestry was started in late 1970s. The concern for social forestry grew out of the increasing shortages of forest products like fuel wood, fodder, and small timber. But as productivity expanded, it gave rise to numerous expectations. Concerns for minor forest products, employment, income, and investment came within its purview. However, economists expressed their concern that from forestry was replacing the traditional crops. According to the scholar social forestry has a tremendous potential.

In an extensive study of three social forestry projects in Orissa undertaken by Acharya (1998) revealed that benefits (both direct and indirect) generated by theses projects were immense by using social cost benefit analysis (SCBA). The study also found that Gopalpur Social forestry Project is the most socially viable project among the three studied. The one of the findings of the study suggested that the community participation in managing the projects were relative effective and sustainable in the project site where the cast homogeneity was of high degree in contrast tot the area of less homogeneity in castes. Moreover, the study suggested that peoples or community participation would be very much required for the success of any social forestry project.

Singh *et al.* (2000) has made an attempt to assess the impacts of social forestry projects on the various stakeholders of the society with reference to hills in Uttar Pradesh using Cost Benefit Analysis. It is found in the study that the three types of plantations namely, bhimal, chir and bhil plus chir, were found to be financially feasible at both 5 and 20 per cent discount rates in terms of both the gross benefit ratio (GBSR) and net present value (NPV) criteria. The bhimal plantation was financially feasible proposition even at the 15 per cent discount rate and of three types of plantation; it had the highest paying capacity. A mixed plantation of banal plus chir is recommended for the community land in the catchments. This would provide benefits from bhimal trees in the short-run and from chir trees in the long-turn. Of three types of plantation, it had the highest NPV and highest GBCR at the 5 per cent discount but the lowest NPV at the 10 per cent discount rate. At 15 per cent discount rate, the NPV was negative. The average annual increases in per capita net income at 5 per cent discount rate ranged from Rs.180 from bhimal plantation in situation C to Rs.458 from chir plantation in situation A. At 10 per cent discount rate, the range of increase in net income was between Rs 45 from chir to Rs. 145 from bhimal plantation. Annual increases in per capita net income of this order represent substantial increase in the total per capita income in the catchment.

In a similar kind of study undertaken by Ninan and Jyothis (2004) found that social forestry projects in Karnataka are not only economically viable and but also socially desirable while using social cost benefits analysis. Applying rigorous tests and sensitive analysis, the study notes that at full benefits, net of all costs including the opportunity cost of grazing benefits forgone, the project reports high profits with the internal rate of return (IRR) exceeding 16 per cent.

COMMON PROPERTY RESOURCE INSTITUTIONS AND EQUITY ISSUE

Poverty, property rights and distributive consequences of community based management are increasingly becoming a major subject of debate, discussions and investigations, particularly in issues of local level collective action or community based natural resource management over the past couple of years. One group of scholars advocates that common property resource appropriators can create and sustain the local management institutions that ensure equitable access to, and income from resource management. Furthermore, CPR literature claims that since poor people are heavily dependent on natural resources, they derive higher income and benefits from CPRs. Some scholars, on the other hand, postulate that compare to non-poor, poor people may depend more on common resources, but in absolute term their dependency is lower. Thus, the equity consequences of the common property resource institutions receive increasing importance in the various studies undertaken by several academicians over the years. With this kind of debate pertaining to the CPR management institutions, different scholars review the following studies in brief to highlight the derived results over last couple of years.

In order to study the equity and distributional aspects of joint forest management committees (JFM) Anderson (1995) found that wealthy and influential villager in the control of forest councils are use theses resources for their personal use and poor resources do not gain from this JFM, based on the 180 households in 20 villages of AP by carrying out field survey. In a similar kind of study, Hill and Shields (1998) observed that community incentives in JFM in India are not so clear-cut. The study undertaken by them is based on 130 households in AP, MP, and Rajasthan based on the random sampling. The findings of the study show that the main losers in the JFM are fuel wood head loaders who are often from the poorest subgroup within the village studied.

However, Sreedhar and Bhaskar (2000) have sought to analysis the impacts of Joint Forest Management (JFM) in Anatpur Forest Circle in Andhra Pradesh on forest rejuvenation and livelihoods of participants' households while focusing on equity and sustainability of JFM. The study was conducted covering a sample of 10 per cent of Vana Samrakshana Samithis (VSS) by using a combination of observation and interview methods. It was found in the study that JFM had considerable impact on forest regeneration and on the livelihoods of the participant households encompassing economic, social, political, and environmental domains. The forestry works alone generated as much as Rs. 1.89 lakh person days of additional employment through which a sum of Rs.75.4 lakhs was paid to the members in the form of the wages. Furthermore, some members could benefits by the sale of NTFPs while other could take up subsidiary occupations. Besides these, the VSSs activities paved the way for the emergence of collective and social consciousness among the people, creating a positive attitude towards education, health, family planning and gender equality. In the same way, the VSSs have contributed to the emergence of leadership among men and women at the grass root level. They were also successful in making the people environment friendly. However equality has been built in component of JFM in the sense that over 90 per cent of the expenditure incurred on forest works is being paid to the members in the form of wages. In addition, the income generating programmes (IGPs) taken up by the women groups were only of small scale in nature because of the meager funds at their disposal. These need to be scaled up through the mobilization of outside resources by undertaking capacity building prgrammes. It was also found that the people's involvement was not up to the desired level and effective community participation, flexibility in operation and promotion of IGPs through the women groups need to be ensured for the sustainability of JFM. This calls for immediate attention of the forest department to initiate appropriate measures.

In a study, Kumar (2002) found that it is the rich who gain under the joint forest management (JFM) over 40 years time horizon and poor are the net looser in this process, using social cost benefit analysis in six villages of Ranchi district in Jharkhand state. The study uses 2.93% social discount rate while estimating the future cost and benefits at present value in terms of social preference to different stakeholders. The low SDR is justified in the study that the future flow of forest in perpetuity is not reduced while extracting as much as forest produce as we desire. The 40 years time horizon has been taken in this study keeping in mind the long gestation period of the forestry projects. In addition, the different weights such as 0.58, 0.82, 0.88, 1.34, are attached to different households by stratifying the households into large farmers (more than 4 ha of land), medium farmers (2 – 4 ha), small farmers (1 – 2 ha), marginal farmers (up to 1ha) and landless farmers respectively.

The study has failed to quantify the externalities such as biodiversity conservation, control of environmental pollution, aesthetic and cultural values, etc. while estimating costs and benefits of joint forest management (JFM) forest projects in the studied villages. Moreover, the stratification households into four types considering only the land ownership mentioned above following the classification norms of the World Bank can be appropriate for the is study. The classification should have been made by taking into account other household endowments such as livestock, income from non-land assets, etc so that it could have become more appropriate in this context. Apart from this, the number of classification should have been confined to two: rich households and landless and poor households so that the distribution of equity could have become quite clear.

In a study of two districts of Nepal, Adhikari (2002) made an attempt to find out the distributional consequences of the community based resource management on multiple users and stakeholder groups in the region studied using some econometric techniques. The study was based on the primary data collected from a survey of 309 households from the mid-hills of Nepal where community forestry intervention has been implemented for last two decades. The findings of the study suggested that poorer stakeholders are currently benefiting less from community forestry rather than relatively better of households. In addition, both annual average gross and net income from community forestry is higher for richer and middle wealth households. Econometric results show that some socio-economic variables of the resource-using group place stringent limits on the extent to which certain groups are able to gain access to and benefit from collective action. In other words, income from common property forestry directly associated with household's private endowments.

From the above reviewed studies it is apparent that the debate of distributional outcome of CPR institutions is yet to be settled down properly setting aside the conflicting views in this regard. Therefore the issue needs further empirical investigations and evidence in order to arrive at a final conclusion.

CPRs INSTITUTIONS AND SOCIO-ECONOMIC HETEROGENEITY

The role of heterogeneity in affecting the collective action such ac community based resource management outcomes are strongly debated as an important issue in the context of natural resource management in both theoretical and empirical literature recently. It is generally argued that in the social, ethnic economic and political heterogeneity can create impediments to the very performance of the community based resource management because it can make communication and cooperation difficult which directly affect the cost and benefits of users as a whole. It is still a challenge for the social scientists to examine whether heterogeneity of all kinds affects adversely the collective action such as community based resource management outcomes. There are broadly two schools of thought pertaining to the impacts of heterogeneity on collective action. One school argues that it does while others say, it does not. However, the clear picture relating this case is yet to be come up with concrete evidence. Owing to the significance of the illustrated issue in the light of increasingly concerns expressed over the natural resource management some studies are reviewed by focusing both school of thoughts below.

First of all the issue of the impact of heterogeneity on collective action arose in mid-1960s. It is Olson (1965) who first contended that more inequality might favour the creation of public goods. He explains that inequality (based on endowments, e.g., wealth and land distribution) is good for the management of the common property resources, as these with greater interests may decide to provide the common good even if the poorly endowed group chooses to free ride. Drawing upon recent theoretical advances in the analysis of collective action, Molians (1998) undertook an econometric analysis of the determinants of successful collective action based on a survey of 104 peasant cooperative institutions in Paraguay. His study shows that controlling as much as possible for the specific characteristics of the community and peasant committee, there is a U-shaped relationship between community inequality and cooperative performance. Dayton and Johnson (2000) developed a simple model of individual household's incentives in a community owned irrigation system and found that social heterogeneity, measured by different villages from which the users of a given system are drawn, is consistently and significantly associated with lower levels of infrastructure maintenance.

In a recent study of 18 forests user groups in Nepal, Ostrom and Varughese (2001) found that heterogeneity (location, socio-cultural and wealth disparity) is not a strong predicator of the community resource management outcomes (collective action) using Logit model. Similarly, Somanath *et al.* (2002) examine the effects of village level of heterogeneity in caste and land ownership, and of female membership on the community based forest management in the Indian Central Himalayas. The study found no evidence that caste heterogeneity or female membership of the van panchayats have any effect by using the Logit regression model. But it is found that greater equality in land ownership may enhance community action.

With a view to discuss the role of heterogeneity in the effective outcomes of common property resources(CPRs) Baland and Platteau (1999) theoretically advocate that inequality has an ambiguous impact on the feasibility of the efficient outcome even though better endowed households contribute more to collective action in a voluntary provisions of problem setting whereas in regulated settings inequality tends to reduce the acceptability of available regulatory schemes and ,therefore, make collective action more difficulty.

There are at least three different arguments put forward by scholars why heterogeneity may be inimical to the success of collective action such as community based resource management outcomes. First, according to Alesina, Baqir, and Easterly (1999) suggest that tastes for public goods may differ across groups and this can reduce the willingness to pay for public goods which may not be the kind of that one's own group wants. Second, very often-political scientists argue that social divisions may throw up politicians who seek to provide private benefits to particular groups rather than providing public goods. Third argument states that heterogeneity makes it harder for communities to reach an agreement about the sharing of the benefits or costs of collective actions. Inequality in power, for instance, may mean that equal division would be unacceptable to the powerful, while any other distribution may be subject to conflict. The basis of this argument is that heterogeneity removes a natural focal point for the consensus and, simultaneously makes groups uncertain about other group's preferences, thus making agreement less likely as each group tries to drive a hard bargain, one that may be unacceptable to other groups. This argument calls for the examination of heterogeneity in the context of community resource management.

For instance, let's take the asset heterogeneity for the efficient common property resources managed outcomes here for explanation. As it is know known that the common property institutions are self organized institutions, and set rules pertaining to the access, use and harvesting and conservation of common property resources such as forest, confined to well defined and delineated boundary within a particular area. It is obvious fact that landless and poor households would like to collect fuel woods, food items fodder, etc., frequently from the common property resources by going against the set rules of limited access in order to meet their daily needs for substance of their livestock and livelihood why because they do not have other private resources to collect those mentioned stable requirements whereas the large and rich households easily meet their daily needs of fuel woods, fodder, etc., since they hold enough private assets such as land and other supplementary private property to support their needs without defying the use and access rules . The restricted access rules affect adversely the benefits of the poor households from commons while these don't have any significant effect. Thus, the heterogeneity of asset would certainly have negative impact on the probability of the efficient community based managed resources outcomes by adversely affecting interest of some group of households who are member of the said institutions. However, the topic is still a challenging subject among the scholars to unravel the relation between heterogeneity in its all forms and collective action over a couple of years now in the context of natural resource management.

CPR INSTITUTIONS AND MIGRATION

In most of the developing countries environmental degradation caused by increasingly population growth and higher level of poverty has been manifested in large-scale deforestation and soil erosion. In addition, there has been massive decline and degradation of CPRs over the last couple of decades affecting poor households adversely in rural area where these are considered as the vital resource base in supporting the livelihoods of the poor in manifold ways. Furthermore, it is evident that common property resources constitute a large part of consumption base of the poor in the rural area (Jodha, 1986). Presently, the adverse impacts of the huge or large-scale decline and degradation of CPRs on poor household in rural area are substantial. One of the major outcomes of the degradation of CPRs is massive migration from rural to urban centre, causing lot of problems in the destination point. This kind of migration is called as environmental stress migration (ESM) or push migration, which constitutes substantial percentage of the total migration (Chopra and Gulati, 1998). However, there are few studies, which discuss the impacts of common property institutions on stress migration from rural to urban centre.

A hypothesis developed by Todaro (1980) suggested that migration proceeds in response to rural-urban differences in expected rather than actual wage or earnings comprised one major explanation. Furthermore, it is the probability of getting a job that determined the decision to migrate. In a study, Greenwoods (1971a, 1971b) found that migration was related positively and significantly to destination wage and negatively and significantly to wages at the place of origins by adopting some quantitative techniques in the context of rural migration in India. With a new approach, Chopra and Gulati (1998) have hypothesized that emergence of rural common property institutions has significant influence on the magnitude of migration from rural to urban area in the light of environmental degradation. While discussing this hypothesis they argue that income in rural economy are typically the sum of those obtained from private property resources and those received from common property resources (CPRs). But the impact of recently emerged CPRs institution on migration behaviour through their effects on rural-urban income differentials have, however, little attention from the researchers .A strengthening of common property institutions increases total expected incomes, thereby tending to reduce the propensity to migrate .In other words increased income generating potentials of CPR institutions works as a disincentive to migrate.

The study is based on data of the sample size of 189 households collected through field survey i six villages in Udaipur district of Rajasthan. The impact of CPRs institutions on migration has been assessed by using the OLS, 3LS and Logit model. It is found that there is a negative relation existing between the CPRs institutions and magnitude of migration in the villages studied. The results of the study suggest that proper specification of property rights (effective CPR institutions such as Joint Forest Management Committees) as a consequence of the effective working of NGOs does reduce environmental stress migration. In addition, the household decision to migrate an/or to participate in common property right creation are interrelated and depends upon the levels of asset ownership, degree of dependence on common land and level of education. In a study of 18 VSS covering Anatpur forest circle of Andhra Pradesh, Sreedhar and Bhaskar (2000) suggested that the seasonal stress migration could be contained to the extent of 44.7 per cent over the pre-joint forest management programme situation. The JFM programme provides number of employment opportunities through its various schemes such as plantation, selling of none timber forest products (NTFPs) and community development projects in the villages every year where JFM committees are in operation.

On the contrary, a study undertaken by Kumar (2002) unfolds different story altogether about the rising rural environmental stress migration (ESM) in the protected area of Jharkhand, using the cost benefit analysis to assess the impacts CPR institutions of social forestry projects on the livelihood patterns of rural households. The data used in this study have been obtained from five villages of Ranchi district of Jharkhand through field survey while using the stratified random sampling. It is found that the migration rate in case poor households are high in the area where joint forest management institutions are quite actively involved in the protection, management and development activities of commons since substantial amount of employment opportunities provided by the government and also created within the CPR institutions, are being appropriated by the large households or rich who hold higher executive posts in these CPR institutions.

CONCLUSION

The poverty and population growth are the major causes of degradation and depletion of common property resources in India. The positive correlation between poverty and environmental degradation has been well established in the economic literature. Besides these, the factors such as market forces, policy changes, technological changes and natural calamities also contribute significantly to the degradation and depletion of common property resources in the country. As it is evident from the above review that most of the social forestry programmes under the aegis of common property regime have been socially viable and economically efficient. However, the equity aspect relating to the common property resource management institutions is not empirically settled down. The issue needs to be addressed properly through empirical investigation in the coming days. The impact of socioeconomic heterogeneity on the effective community based common property resource management is not clear and calls for further empirical as well theoretical examination in the country.

The debate on the relation between CPR institutions and environmental stress migration (ESM) calls for elaborate and comprehensive empirical evidence to reach a particular conclusion against the background of the conflicting findings of the scholars over this issue as illustrated above. The scholars working in the natural resource management (NRM) should carefully look into issues and problems pertaining to the impacts of the CPR institutions on equity, environment and livelihood pattern of the stakeholders while carrying out studies using both qualitative and quantitative techniques. It is very evident from the above discussion that the findings of the studies reported vary across the regions in the country due the existence of diversity in the local problems and issues involved in the management systems of CPRs. Though community based management has been considered as one of the best systems while managing common property resources, but it is very difficult to reach a general conclusion that it takes care of equity, environmental sustainability and livelihood requirements of the stakeholders due to the lack of sufficient both empirical and theoretical literature in this field. Therefore, a lot of studies should be carried out in the area of common property resource management institutions both in terms of theory and empirical investigation focusing various aspects of the CPRs in India before reaching a any general conclusion.

NOTES

1. Joint Forest managment Committees carry different name in different states in India. For instance it is Vana Samrakshyan Samity (VSS) in Orissa.
2. The forests of India are a critical resource for the subsistence of rural peoples throughout the country, but especially in hill and mountain areas, both because of their direct provision of food, fuel and fodder and because of their role in stabilising soil and water resources. As these forests have been increasingly felled for commerce and industry, Indian villagers have sought to protect their livelihoods through the Gandhian method of satyagraha non-violent resistance. In the 1970s and 1980s this resistance to the destruction of forests spread throughout India and became organised and known as the Chipko Movement.
3. See, for instance, *The Resolution of June 1990*, Ministry of Environment and Forests, Government of India.

REFERENCES

Adhikari, B. (2002), Property Rights and Natural Resource: Socio-Economic Heterogeneity and Distributional Implications of Common Property Resources and Management, *EEE Working Papers Series-N.3.*

Acharya S. K. (1998), *Economic Evaluation of Various Forestry Schemes: Case Studies from Orissa*, Unpublished Ph.D Thesis, IIT Kanpur, U.P., India.

Agrawal, A. (2001), Common Property Institutions and Sustainable Governance of Resources, *World Development*, 10(29): 1649-1672.

Alsina, Alberto, Reza Baqqir, and William Easterly (1999), Public Goods and Ethnic Divisions, *Quarterly Journal of Economics*, 44(4): 1243-84.

Anderson K. E. (1995), Institutional Flaws of Collective Forest Management, *Ambio*, 24 (6): 349-53.

Baland J. M and J. P. Plateau (1996), *Halting Degradation of Natural Resources: Is there a Role for Rural Communities?*, Clarendon Press: Oxford.

Baland J. M and J. P. Plateau (1997), Wealth Inequality and Efficiency in the Commons Part-I: The Unregulated Case, *Oxford Economic Papers*, 49: 451-482.

Baland, J. and J. P. Platteau (1999), The Ambiguous Impact of Inequality on Local Resource Management, *World Development*, 5(27): 773-788.

Bardhan, P., Jean-Marie Baland, Sanghamitra Das, Dilip Mookherjee and Rinki Sarkar (2002), The Environmental Impact of Poverty: Evidence from

Firewood Collection in Rural Nepal, http://www.bu.edu/econ/ied/dp/papers/dp126.pdf

Barbier E. B. (1997a), The Economic Determinates of Land Degradation in Developing Countries, *Philosophical Transactions of the Royal Society*, 352: 891-899

Barbier E. B. (1997b), Introduction to the Environmental Kuznets Curve (Special Issue), Environment and Development Economics, 2(4): 369-380.

Bromley D. W. (1991), *Environment and Economy: Property Rights and Public Policy*, Oxford University Press: Oxford, UK.

Chopra, K. C. Kadekodi, and Murthy M. (1989), People's Participation and Common Property Resources. *Economic and Political Weekly*, (24): A189-A195.

Chopra, K. and S. C. Gulati. (1998), Environmental Degradation, Property Rights and Population Movements: Hypotheses and Evidence From Rajasthan (India), *Environment and Development Economics*, 3(1): 335-57.

Ciriacy-Wantrup and Bishop (1975), Common Property as a Concept in Natural Resource Policy, *Natural Resource Journal*, (15): 713-727.

Dayton-Johnson, J. (2000), Determinants of Collective Action on the Local Commons: A Model with Evidence from Mexico, *Journal of Development Economics*, (62): 181-208.

Greenwood, M. (1971a), An Analysis of the Determinants of Internal Labour Mobility in India, *Annals of Regional Science*, 5(1): 137-151.

Greenwood, M. (1971b), A Regression Analysis of Urban Areas of Less Developed Country: The Case of India, Journal of Regional Science, 2(11): 253-262.

Grossman, G. and A. Krueger (1995), Economic Growth and the Environment, *Quarterly Journal of Economics*, (11): 353-377.

Hardin, G. (1968), The Tragedy of Commons, *Science*, 162: 1243-1248.

Heltberg, R. (2001), Determinants and Impact of Local Institutions for Common Resource Management, *Environment and Development Economics*, (6): 183-208.

Hill, I. and Shield, D. (1996), Incentives for Joint Forest Management in India: Analytical Methods and Case Studies (Technical Paper No.394), World Bank: Washington D.C.

Iyenger, S. (1989), Common Property Resources in Gujarat: Some Findings about their Size, Status and Use, *Economic and Political Weekly*, June 24 : A67-A77.

Jalal, K. F. (1993), Sustainable Development, Environment and Poverty Nexus, *Asian Development Bank Economics and Development Resource Center Occasional Papers*, No.7, Manila.

Jodha, N. S. (1986), Common Property Resources and the Rural Poor in Dry Regions of India, *Economic and Political Weekly*, 27(21): 247-264.

Jodha, N. S. (1995), Studying Common Property Resources: Biography of a Research Project, *Economic and Political Weekly*, XXX: 556-559, 11 March.

Kumar , S. (2002), Does Participation in Common Pool Resources Management Help the Poor? A Social Cost Benefit Analysis of Joint Forest Management in Jharkhand, India, *World Development*, 5(30): 763-82.

Lopez, R. (1998), Where Development Can or Can't Go: The Role of Poverty and Environment Linkages, Annual World Bank Conferences in Development Economics , B. Pleskovics and J. Stiglitz (ed.), World Bank: Washington D.C.

Molians , J. R. (1998), The Impact of Inequality, Gender, External Assistance, and Social Capital on Local Level Collective Action, *World Development*, 26(3): 413-490.

Government of India (1988), National *Forest Policy 1988*, Ministry of Environment and Forests, New Delhi, http://envfor.nic.in/welcome.html

Naik, G. (1997), Joint Forest Management: Factors Influencing Household Participation, *Economic and Political Weekly*, Nov-29: 3084-3089.

Ninan, K. N. and S. Jyothis (2004), Social Forestry: A Case Study from Karnataka, in Gopal K. Kadekodi (eds.), *Environmental Economics in Practice: Case Studies from India*, Oxford Press : New Delhi.

Olson, M. (1965), The Logic of Collective Action: Public Goods and the Theory of Groups, Harvard University Press: Cambridge.

Ostrom, E. (1990), Governing the Commons: The Evolution of Institutions for Collective Action, Cambridge University Press: Cambridge.

Ostrom, E. and Varughese G. (2001), The Contested Role of Heterogeneity in Collective Action: Some Evidence from Community Forestry in Nepal, *World Development*, 5(29): 747-765.

Pasha, S. A. (1992), Common Pool Resources and the Rural Poor: A Micro Level Analysis, *Economic and Political Weekly*, 27(46): 2499–2503.

Patnaik , L. K. and A. Bharthuar (1991), Ecological Benefits of Five-Years Old Social Forestry Village Wood Lots in Orissa, *The Indian Forester*.

Saxena, N. C. (2000), Research Issues in Forestry India, *Indian Journal of Agricultural Economics*, (23): 2629-2630.

Reddy ,V. R. , Baghirath Behera and D. Mohan Rao (2001), Forest Degradation in India: Extent and Determinants, *Indian Journal of Agricultural Economics*, 4(56).

Somanath E., R Prabhakar, and Bhupendra Singh Mehta (2002), Collective Action for Forest Conservation: Does Heterogeneity Matter? http://discuss.santafe.edu/files/sustainability/sustain09.pdf

Shukla, Rohit (1992), Social Forestry in Gujarat Employment and Income, *The Price of Forest*, Anil Agarwal (ed.) Proceedings of a Seminar of Centre for Science and Environment: New Delhi.

Singh R. B., Y. S. Chauhan, S. R. Yadav and S. C. Verma (2000), An Economic Analysis of Afforestation of Community Lands in Uttar Pradesh Hills, *Indian Journal Of Agricultural Economics*, 3(55): 444-445

Sreedhar G. and K. Bhaskar (2000), Impacts of Joint Forest Management on Forest Rejuvenation and Livelihoods of Participant Households: Micro Level Experiences from Andhra Pradesh, *Indian Journal of Agricultural Economics*, 3(55): 456-457.

Forest Survey of India (1999), *State of Forest Report 1999*, Dehradun.

Todaro, M. P. (1980), Internal Migration of the Developing Countries: A Survey, in A Easterlin, (ed.), *Population and Economic Change in Developing Countries*, University of Chicago Press: Cambridge.

Wade, R. (1987), The Management of Common Property Resources: Collective Action as an Alternative to Privatization or State Regulation, *Cambridge Journal of Economics*, (11): 95-106.

23

PRIVATISING THE COMMONS: A CRITICAL REVIEW OF THE PROPERTY RIGHTS PARADIGM

Zakir Husain and Dr. Rabindra Nath Bhattacharya

INTRODUCTION

Common Pool Resources (CPR) refers to natural resources jointly used by a large number of people with co-equal rights to the resource. Unlike public goods, however, their consumption is rival - exploitation of the resource by one individual reduces the stock of resource, thereby affecting consumption by others. In other words, consumption of CPRs generates *strategieic externalities* in the shape of reduced consumption – either in the present, or (which is more common) in the future. This problem is serious, as the cost of applying the exclusion principle may be prohibitive. Examples of such resources consist of gleamings from harvests, ground water aquifer, fishing stock, the ozone layer, etc. Such resources may be *local*, i.e. used by local communities (forests); or *global*, i.e. affecting several countries (ozone layer). This study focuses on local commons.

CPRs contribute to rural welfare in many ways. They constitute an important source of stable consumption, add variety to diet, provide low cost inputs to the production process, and contribute significantly to rural income (specially of the poor households) in various ways. They are also a source of employment, particularly amongst women, children and elderly persons. CPRs also form part of the social security system cushioning the rural poor from the effects of seasonal declines in income. Finally, they constitute an important component of the rural resource base. Their destruction leads not only to the loss of an important part of this base, but may affect the functional stability (resilience) of the entire resource base. It is important, therefore, to both conserve and nurture CPRs.

Empirical studies have shown that the area under CPRs is declining over time (Jodha, 1986). While some economists would attribute this to degradation through over-use, this would be a mistake. Such studies have also shown that privatisation is an important means of destroying common rights over natural resources. It is necessary to analyse this trend and examine its implications for the sustainability of the rural resource base. In the present study we shall review the existing literature on this issue. We will attempt to establish Common Pool Resources as an important form of property rights that both increases rural welfare, specially for the rural poor, and ensures the sustainability of the resource system.

PRIVATISATION OF NATURAL RESOURCES

Tragedy of the Commons

Economists have traditionally believed that CPRs are inevitably over-exploited and exhausted – the Tragedy of the Commons. The origin of this problem lies in the inability of the property rights system to ensure that the consequences of the actions of resource owners are restricted to them, leading to over-utilisation of the resource. Various authors have stated this: Aristotles (Jowett, 1967), Marcett (in Bromley, 1989), Lloyd (in Edney, 1980), Warming (Anderson, 1911), Gordon (1954), Scott (1955), and Cheung (1971). The most influential statement of this proposition came from Hardin (1968), who also coined the term 'Tragedy of the Commons'.

Hardin's prediction of tragedy in the commons was based upon an analysis of the open field system prevailing in medieval England. Grazing land was commonly owned by the villagers and formed a stable system initially. But, with demographic expansion, pressure on the commons mounted and the rate of exploitation exceeded the rate of replenishment.

The problem here is that the institutional arrangement - or, to be more specific, *the property rights system* - is not able to prevent the benefits of an individual resource user's environmentally conscious behaviour from dissipating to other (short-sighted) resource users. (Neither are the costs of environmentally degrading action restricted to the offender.)

This suggests that a solution to the tragedy of the commons lies in altering the property rights system. Early writers on this subject believed that this modification must be directed towards a property system that would facilitate internalising the costs and benefits of resource use. This could best be achieved, they argued, under a system where private property rights prevailed. For instance, Hardin stated, "the tragedy of the commons as a food basket is averted by private property or something formally like it." (Hardin, 1968).

The argument in favour of private property rights has been put forward from three different angles: from the viewpoint of economists, from the viewpoint of zoologists, and from the viewpoint of behavioural psychologists. We shall first examine the proposition by economists.

Economist's Viewpoint – Demsetz and Posner

The chief proponent, among economists, of a system of private property rights has been Harold Demsetz. His analysis has it's theoretical foundations in Coase's classic article "The Problem of Social Cost" (Coase, 1960). Coase had argued that the problem of externalities could be solved if the rights of the various parties are well defined. This would enable the parties to take into account all the costs of their activities.

Based on this argument Demsetz (Demsetz, 1966 and 1967; Alchain and Demsetz, 1975) espoused the desirability of a system of private property rights to prevent over-use and depletion of natural resources.

> "If the main allocative function of property rights is the internalisation of beneficial and harmful effects, then the emergence of property rights can be understood best by their association with the emergence of new or different and harmful effects.'
>
> 'Changes in knowledge result in changes in production functions, market values, and aspirations. New techniques, new ways of doing the same things – all invoke harmful and beneficial effects to which society has not been accustomed. ... (T)he emergence of new property rights takes place in response to the desires of the interacting persons for adjustment to new benefit-cost possibilities.'
>
> '... property rights develop to internalise externalities when the gains of internalisation becomes larger than the costs of internalisation. Increased internalisation, in the main, results from changes in economic values, changes which stem from the development of new technology and the opening of new markets, changes from which old property rights are poorly attuned." (Alchian and Demsetz, 1975)

The crux of Demsetz' position was that private property is the most appropriate way to make the individuals internalise the externalities. In his words, "A primary function of property rights is that of guiding incentives to achieve a greater internalisation of externalities" (Demsetz, 1967). In contrast, under communal ownership, the costs associated with the use of that particular resource does not fall on the users. Therefore, the resource will be over-used, as the person seeking to maximise the present value from his use of the resource will overlook that part of costs that is borne by others. Negotiation to curtail excessive exploitation of the resource is of course possible, but the costs of negotiating between the large number of resource users and the costs of policing the agreement will be substantial.

Demsetz proceeds to argue that an individual owning land maximises it's present value by taking into account alternative future streams of benefits and costs, and choosing that stream which maximises the present value of his land. This involves a consideration of both present and future conditions. In effect, therefore, he acts as a broker between the present and future generations. But in the case of communal property there is no mechanism to ensure that the claims of the future generation will be given their due weightage, resulting in the present generation being given excessive preference. In other words, the impact of an individual's actions on the future generation – as well as on other members of the present generation – are not taken into account fully in the case of common property resources.

The incentive effect of a private property regime had long been recognised (though not explicitly stated), as is attested by the following excerpt from Lloyd: "The common reasons for the establishment of private property in land are deduced from the necessity of offering to individuals sufficient motives for cultivating the ground, and of preventing the wasteful destruction of the immature products of the Earth" (in Balland and Platteau, 1996).

Hardin, too, had recognised the superiority of private property rights.

> "An alternative to the commons need not be perfectly just to be preferable. With real estate and other material goods, the alternative we have chosen is the institution of private property coupled with legal inheritance.... we are not convinced, at the moment, that anyone has invented a better system. The alternative of the commons is too horrifying to contemplate. Injustice is preferable to total win." (Hardin, 1968).

Posner (1977) has expressed similar views regarding the ultimate wisdom of institutional change in the direction of atomistic property arrangements over resources. According to him, the proper incentives for economic efficiency were created by the parcelling out of mutually exclusive rights to the exclusive use of particular resources among the members of society. If every piece of land is owned by someone, in the sense that there is always an individual who can exclude all others from access to any given area, then individuals will endeavour by cultivation or other improvements to maximise the value of land.

He argues that there are three criteria of an efficient system of property rights:

1. Universality*:* All resources should be owned by someone;
2. *Excl*usivity: Rights to use the resource belong exclusively to the owner; and,
3. Transferability: Such rights are transferable to some other person on a temporary or permanent basis.

These characteristics are satisfied only by a system of private property rights.

de Alessi (1980) also provides a coherent critique of communal property rights. He argues that common ownership will imply reduced investment, increased entry to capture the rent, and earlier harvesting of the resource. The recognition of these problems often leads to restrictions on the use of privately owned capital, leading to the predominance of labour-intensive techniques. Consequently, the average and marginal productivity of labour will be lower, and those of capital higher, than under private ownership. Moreover, since common ownership discourages the postponement of capture (as the stock may be appropriated by someone else), the quantity of resource appropriated will vary more widely over time so that the resource is more likely to become exhausted. The choice of resource use may also be affected. For instance, instead of planting a crop with a higher present value but a longer gestation lag, individuals have the incentive to grow faster growing crops, or grazing cattle - activities on which he incurs lower transaction costs.

Biologist's Viewpoint – Wynne-Edwards and Ardrey

Privatisation, we pointed out before, has also been supported by contributors from other fields. Wynne-Edwards' (Wynne-Edwards 1962 and 1965) ethnological spatial dispersion theory, for instance, deals with the management of common resources by sub-human animals. The basic tenet of this theory is that animal communities self-regulate their population to avoid over-exploitation of food sources through a system of territoriality.

Members of the consuming community – usually the males - compete for a limited number of territories in the resource bearing area, and since these territories are large, winners disperse sparsely over the resource, preventing over-exploitation. Females generally mate with territory winners, so that the latter's "successful" genes are passed onto the next generation. Subsequently, the next generation is not only genetically superior, but also smaller in size.

Territoriality, therefore, performs several functions: it helps in genetic selection, reduces the rate of resource consumption, generates a social hierarchy, with territory winners forming the elite, and prevents direct competition for resources that could otherwise destroy the resource pool.

It is doubtful to what extent Wynne-Edwards's theory is applicable to human societies. For one, human territories do not display the same functions as it does for sub-human animals. Besides, loosers are not treated as expendables in human society – specially with the emergence of the welfare state. Nevertheless, Ardrey (1966 and 1970) has tried to extend Wynne-Edwards's theory to human society. He proposed that human societies are governed by competitive principles and that the aggressive social behaviour of human beings is based on innate and unalterable biological factors. He further points out that there is co-existence of equality of opportunity with genetic inequality of ability. This creates biological divisions within a society, forming the basis of the process of genetic selection. Simultaneously, there is inter-dependence between members of the community, so that the survival of the community calls for some self-sacrifice on the part of its members.

In contrast to Wynne-Edwards, Ardrey stressed that competition among humans often transcends conventional forms. Further, the recognition of the inherent biological equality is psychologically manifested as a basic drive for identity. Achieving control of the resource distribution can satisfy this drive. This is possible either by achieving control of the resource distribution. This is possible either by occupying a place in the community hierarchy, or through possession of a territory. Thus, according to Ardrey, the function of a territory is to demarcate the owner's biological superiority and obtain for him the secure control of a portion of the resource stock of the community.

Psychologists Viewpoint – Crowe and Edney

Psychologists have also suggested the utilisation of the concept of territories to resolve the commons dilemma. Crowe (1969) suggested that large territories should be divided into small homogeneously valued groups.

Territorialising implies, in this context, the division of the common resource into sub-divisions with individual (or sub-group) responsibility for each. Territorialising does not simply mean the creation of boundaries (and their maintenance); it is a system of behaviour that has cognitive, temporal and spatial aspects. The essence of this strategy, however, is the division of the whole resource in such a way that individuals are given the freedom to utilise their allotted portion, bur depletion of that portion minimally affects other portions. Thus, the solution is to tackle directly the problem by dismantling the commons itself.

Edney (1980) has argued that such a strategy is acceptable for the following reasons:

(a) The individual is given the freedom to utilise the resource allotted to him in a manner which he believes to be rational;

(b) The individual is also given the responsibility to maintain the division;

(c) It minimises competition over the resource and allows territory holders a sense of individuality and control;

(d) It relaxes the individual's responsibility for the group, and vice versa. Thus territory holders are not required to practice altruism, reducing the scope for free-riding; and,

(e) Need for external governance and maintenance of the territory is reduced.

Results of Experimental Studies

Support for the beneficial effects of privatisation also comes from an experimental study by Cass and Edney (1978). They have argued that the division of the resource pool into distinct territories will make resource management more efficient, with higher *individual* exploitation rates, even though each individual territory is smaller and has lower carrying capacities than the common pool. This is because:

1. The allotment of distinct territories will develop a greater sense of individual responsibility in each harvester, reducing responsibility diffusion for a crisis and discouraging individual harvesters from thinking that they free ride on others;
2. Such a system reduces the uncertainties involved more 'controllable'. Division of the resource pool into individual territories eliminates the problem of free riding and reduces the problem to one of self-control or self-management. Further, with the eradication of external effects, property owners become solely responsible for the outcome of their actions, as well as being able to predict these consequences with greater reliability.

The experiment studied the effects of making the impact of resource consumption visible to the resource community and dividing the resource pool into territories, one for each harvester. Both were expected to improve resource management and allow exploitation at higher rates in the long run, but the effect of each factor was expected to be dependent on the other. The experiment was set up as a commons dilemma situation, in which the subjects play in groups of four, 'exploiting' a regenerating computer controlled resource pool. Pay-offs to individual harvesters at the end of the experiment ware based on total points accumulated, with subjects being able to choose to extract points from the resource stock rapidly or slowly over a number of rounds. The dilemma faced by these resource users was whether to extract points rapidly for themselves, or to limit extraction in each round allowing time for the community pool to regenerate. A game in such simulation exercise continued till the pool was depleted, or a time limit was reached.

The results of the experiments suggested that both visibility of resources and territorialisation contribute to efficient resource management. A combination of both these conditions resulted in a near optimum strategy being adopted; on the other hand, allocation of territories without resource visibility resulted in under exploitation out of caution. The authors argued that private territories increased the predictability of the system and focused responsibility on the owner, thereby creating a greater sense of control. This led to the adoption of conservative exploitation strategies. The availability of feedbacks about the impact of such strategies – in the form of resource visibility – allows resource users to shed their over-cautiousness and move to a strategy closer to the optimum.

Examples of Sustainable Privatised Resources

Advocates of privatisation have also cited the emergence of successful private property rights in the case of endangered resources.

Demsetz's discussion of the relation between the advent of commercial fur trade and property rights in the Montaigne Indian society is one such example (1967). Before the establishment of the fur trade, game was communally owned and hunting was carried out to satisfy food and clothing requirements. Externalities were undoubtedly present, but the scale of hunting was such that they were not significant. The advent of the fur trade had two effects: firstly, the value of fur was considerably enhanced; secondly, in response to the first development, hunting intensity increased sharply. Both these effects enhanced the externalities associated with hunting to significant levels. This led to a change in the system of property rights, with the onset of private ownership. The first stage witnessed the temporary allotment of land, this was followed by seasonal allotment of land, and finally full-fledged hunting territories earmarked for groups.

In addition, Demsetz notes several other cases of successful privatisation: for the Indians of the Labrador Peninsula and of the North-West Coast (1967). Harper-Fender (1981), too, notes the favourable impact of privatisation of natural resources to conserve the fragile supply of fur-bearing animals amongst Cree Indians of Saskatchewan.

A CRITICAL REVIEW OF PRIVATISATION

Early writings on the commons were characterised by the following features. Firstly, they all believed that CPRs were over-utilised and would therefore be depleted. Secondly, the root of the problem lay in the system of property rights, which prevented the consequences of resource utilisation accruing to individual resource users. Thirdly, the emergence of distinct property rights in the resource would lead to resource conservation. Hence, CPRs should be privatised.

This view, however, overlooks several points:

1. Establishment and maintenance of private property rights has costs, which may be quite substantial.
2. The system of property rights prevailing at a point of time depends upon the economic surplus being generated.
3. Even privately owned resources can be over-exploited under certain conditions.
4. Community-based management of resources may also provide a viable sustainable alternative.

The subsequent discussion will elaborate these points.

Costs of Property Rights

It has been argued that the establishment of property rights has its costs and benefits. Whether or not property rights will be established, the form they will take, etc. will depend upon their Marginal Cost and Marginal Benefit curves. Consequently, as there is a change in these curves, there will be a corresponding change in the property rights arrangement.

For instance, when the Marginal Benefit curve is MB, and the Marginal Cost is MC, property rights will not be established. But a change in technology, for instance, may result in the costs of enforcing property rights falling to MC*.

For instance, private property in land in the American Great Plains did not develop fully before the 1870s because the cost of fencing was too high, making it impossible to prevent livestock from crossing range boundaries. But, the use of barbed wire as fences greatly reduced the costs of fencing and led to the demarcation of property rights after 1870.

This implies that the costs of establishing property rights are important and we cannot claim beforehand that private property rights are efficient under any situation. Such costs occur under the following heads:

1. Costs of establishing and protecting property rights;
2. Costs of decision making with respect to use of a scarce resource;
3. Costs of establishing institutions to negotiate between agents; and,
4. Costs of policing after the optimal level of resource use has been determined.

For instance the person owning the resource has to be clearly agreed upon, and non-owners excluded. With regard to CPRs it is not simply enough to demarcate boundaries between tenements and assign ownership rights to them – though this itself is a costly operation. It is also necessary to protect each owner from intrusion by others. Alternatively, if the decision were to retain communal management, there would be no such costs as long as membership of the resource community can be controlled. In medieval England, for instance, this was ensured through the system of frankpledge. The transaction costs of establishing and operating a private property regime are, therefore, enormous.

Property Rights and Economic Surplus

The property rights school argues that private property rights will maximise the economic surplus generated. This is based on the assumption that the system of property rights prevailing determines the economic surplus. Bromley argues, however, that the opposite is true – it is the magnitude of economic surplus being obtained that will determine the appropriateness of the system of property rights. He is, therefore, reversing the functional relation:

$$\text{Surplus} = f\,(\text{Property Rights})$$

to:

$$\text{Property Rights} = g\,(\text{Surplus}).$$

Dahlman (1980), for instance, has shown that the open field system operating in medieval England was superior to a system of private property rights. He pointed out that the optimal scale of operations was different for arâble and pastoral activities. While crop cultivation could be carried out in small discontinuous plots, grazing had to be carried over a wide area. This led to differing types of property rights emerging for these two activities. In the former private plots were prevalent. But, for pastoral activities, division of the land and the resultant restrictions on access would require frequent negotiations to determine the terms and conditions on which cattle could be grazed on some other's land. In addition, policing would be necessary to detect over-grazing, or breaking any other terms of the agreement between the parties. These two conditions led to high transaction costs. The problem was by-passed by the introduction of open access to pastoral land. Simultaneously, the right to withdraw land from communal use was withdrawn – so that the ability of individuals to threaten others was reduced. Economic pressures (grazing could be carried out only in large consolidated holdings) also contributed the stability of the system.

Sustainability and Privatisation

Establishing private ownership would not, however, necessarily eliminate the danger of resource depletion. The reason is that the rate of exploitation would depend upon private costs and benefits from extraction. In the presence of externalities there would be a divergence between such private costs-benefits and social costs-benefits leading to a solution that is not sustainable in the long run. Such divergence's may occur of the cost of exploiting the resource is low, the gross returns from exploiting the resource is high, there is insecurity regarding the ownership of the resource, or if future income is discounted at a very high rate.

So over-exploitation of a privately owned resource, when the objective is to maximise individual present value maximisation, is quite possible as shown in various mathematical models (for a list of citations, see Larson & Bromley, 1990). In fact over-exploitation may be the optimal strategy under the following conditions:

1. When the minimum viable resource stock is positive, extinction is optimal if the initial resource stock is small;
2. When the minimum viable resource stock is zero and the discount rate exceeds the growth potential of the species (Cropper, 1988).

While depletion may not result from present value maximisation if harvesting costs increases as the stock decreases. The resulting stock could still fall to such a low level that it has become depleted in an 'economical' – if not 'physical' – sense (Larson and Bromley, 1990).

So we see that ownership is not the crucial element in determining whether a resource is used sustainably or not – it is control over possible uses of the resource.

Environmental Degradation, Property Rights and Attenuation

In this context, the meaning of property rights becomes important. Property rights refer to:

> "the sanctioned behavioural relations among men that arise from the existence of things and pertain to their use. Property rights assignments specify the norms of behaviour with respect to things that each and every person must observe in his interactions with other persons, or bear the cost of non-observance. The prevailing system of property rights in a community can be described, then, as a set of social and economic relations denoting the position of each and every individual with respect to the utilisation of scarce resources." (Furubotn and Pejovich, 1972)

This has two implications. Firstly, the relation is not between a person and the resource; it is a relationship amongst individuals with respect to resources. Secondly, ownership of property rights does not imply unfettered enjoyment of the benefits from the property. Rather it implies that the prevailing institutional structure will specify the rights and privileges of the owner with respect to the flow of benefits accruing from the property.

Burrows (1979) observes:

> "One of the important functions of the legal system is to define the legitimate uses of property. Even in a social system based on private ownership of property, society, through the legal system, confers on the owners only limited rights of use. ... It is clear, therefore, that the ownership of property is essentially the ownership of rights to some, but not all, of the services that property is capable of yielding. In addition to allowing the owner to enjoy these services, private ownership normally provides for the exclusion of others from also enjoying the permitted activities. ..."

Now environmental crises occur because resource owners do not give sufficient weightage to the welfare of future generations. So they will undertake activities, or use their property right, in a manner that will harm future generations. Under the prevailing institutional set up, or rather under the prevailing frame of rights, such activities may not be considered illegal and can be continued. It is only when the existing structure of rights to the future income stream from the resource is curtailed in a manner by which the welfare of future generations are given sufficient weightage that the environmental problem can be solved. Such restrictions on owned resources are referred to as attenuation: "Any limitation on the way in which property rights may be used constitutes attenuation." (Quiggin, 1988).

The problem, therefore, is not due to the form of ownership prevailing, but in the specification of the rights enjoyed by the property owners: "Common property of natural resources is no more a tragedy in terms of environmental depletion than private property. It all depends on what social institutions ... are guiding resource use in either case." (Ciriacy-Wantrup, 1971).

Open Access versus Community Management

The property rights school, however, bypassed this issue by equating common property with open access conditions. The property rights school has held that either property is private, or it is held by the state. According to this view, resources not amenable to private appropriation are 'common property'. This implies that such resources are not owned by anyone: it is a free good. This approach view CPRs as open access resources freely available to all users.

It has been argued by Ciriacy-Wantrup and Bishop that this usage is "at odds with the long-standing meaning of the concept ...."(1975). They further point out that:

> "The meaning of the concept "common property" is well established in formal institutions such as the Anglo-Saxon common law, the German land law, the Roman law and their successors. It is well established in informal, institutional arrangements based on custom, tradition, kinship and mores..... economists are not free to use the concept "common property resources" or commons" under conditions where no institutional arrangements exist. Common property is not "everybody's property" ... to describe unowned resources (res nullius) as common property (res communes) as many economists have done for years in the case of high seas fisheries, is a self-contradiction" (Ciriacy-Wantrup and Bishop, 1975).

Many CPRs (including Hardin's common field system for a long period, for instance) are communally owned and managed sustainably. Resource users may realise the benefits of co-operation and create institutions (both formal, as well as informal), which incorporates ecological conservation as an objective within its structure. Such institutions contain mechanisms for both exclusion and allocation of the resource. Examples of such community-based resource management systems includes the Spanish Huertas, the Zanjiras in the Phillipines, the Alpine grazing systems, Japanese common land system, the Lofoten fishery system, etc. (Berkes, 1989; Ostrom, 1990). This has led to reclassification of the different property rights system into the following divisions:

The property rights school equates open access resources with CPRs. But community based systems can also provide a sustainable alternative by building up group identity. Group identity encourages co-operation because members of a social group tend to regard other members favourably and believe them to be trustworthy, honest and co-operative.

Other members of the same group are expected, therefore, to reciprocate co-operative behaviour. In addition, inclusion within a social group reduces social distance between members so that they make less distinction between their own and other's welfare. In fact, experimental studies have indicated that feed backs regarding group outcome has a greater impact than feedback on his individual performance (Kramer and Brewer, 1980).

More importantly, a community-based management system can minimise the substantial transaction costs associated with establishing and operating a resource regime. In an *n*-person resource community ${}^{n}C_2 = [n(n-1)/2]$ agreements would be necessary. Naturally, the costs of such an extensive negotiation process would be significant. An institutional alternative to the series of bilateral exchanges is to create a CP regime that internalises the multiple negative externalities. For instance, instead of having numerous individual owners of upland forestland, a CP regime can be devised where the community earns more from down hill land, which compensates them for not cutting timber up hill.

As McKean (1996) argues:

> "CP regimes may become desirable when more intensive resource use multiplies the externalities between parcels, or when individuals face insurmountable transaction and information costs in Coasean contracting. ... There is probably some point at which economies of scale in negotiating takeover, and collective decision making, collective agreement on fairly restrictive use rules, and collective enforcement of those rules becomes easier (less time, lower transaction costs for the owners) than endless one-on-one deals."

This becomes crucial in the context of low value resources.

In addition, CP rights can serve to operate as a way of privatising rights to resources without sub-dividing it. Natural resource systems are complex – their optimal scale of operation may be large. In that case, they are more productive when left intact.

The optimal scale may be large and may constitute the entire system for several reasons. For instance, the CPR may be physically indivisible or may lack distinct boundaries so that division of the resource may be technically impossible. The resource may be mobile and variable, or even unpredictable, so that resource users may prefer to share the entire area, and decide where to concentrate use at a particular time, rather than parcel the system and impose risks on some individuals. Sometimes the administrative infrastructure that is necessary to support individual property rights may not exist – cheap fencing material or courts to resolve disputes may be absent. In that case, sharing rights is an economical way of ensuring enforcement of the group's right.

Of course, it is possible for some institutional rule like Majority Decision Rule to emerge when there is a large number of bilateral bargaining. Since such arrangements are also necessary under communal ownership there appears little to choose between these two forms of ownership in this regard. However, this overlooks the voluntary nature of such arrangements. In the case of private property rights, the emergence of such an arrangement creates scope for appropriating a more than fair share by threatening to withdraw from the organisation. In the case of communal ownership, however, individuals are bereft of such bargaining power so that establishment of institutional arrangements and their preservation becomes easy. Secondly, the entire community carries out monitoring. This spreads out the costs of monitoring over the entire community, so that per head contribution is low. Users monitor each other's; they also band together to patrol the entire resource system to prevent outsiders from exploiting the resource. Rotation rules place the two actors most concerned with cheating in direct contact with each other so that the fear of being cheated leads to monitoring. Pro-social norms and conventions are employed. This allows the community to economise on monitoring costs.

Finally, vesting resource rights in those living in proximity to the resource gives them incentive to monitor the resource properly. People living close to the resource have certain strategic advantages over others in the sense that they have the potential to exploit and degrade the resource more easily. They can also monitor resource use more easily. Vesting property rights in these people strengthens the incentive of those most capable of enforcing rules about access to the resource to design rules that protect and enhance it. Vesting those rights in others instead destroys those incentives and converts those nearest the resource into likely poachers. Any rights scheme needs to consider the strategic importance of vesting resource rights in the communities living closest to the resource.

CONCLUSION

To conclude, private property rights regimes are not sufficient to ensure environmental sustainability – they are only necessary conditions. What is necessary is that rules exist for exclusion of 'outsiders' and for resource sharing. This implies that no single type of property rights can be prescribed a priori as a remedy for environmental degradation. However, community-based resource management systems have the following advantages:

1. They utilise local skills and resources.
2. They are compatible with local culture.
3. Information requirements are less and can be obtained more easily.
4. Monitoring costs are lower as user participation increases management legitimacy and leads to better compliance.

5. They ensure equity in resource allocation.
6. They enable the poor households to access the resource.

Consequently, the gradual privatisation of CPRs found in various studies (Jodha, 1986; and Pasha, 1992) is not encouraging. In fact, it is the shrinkage in the area under common lands that has contributed to the increasing pressure on such resources leading to their degradation (Sen, 1981). It is, therefore, necessary to reverse this trend and encourage the establishment of decentralised management structures.

Table 1
Accrual of Costs/Benefit to Resource User and Community

Action	*Increase in Output*	*Over-grazing*
Addition of extra cattle	Benefit monopolised by individual	Cost of depleted common land spread across all herdsmen
Not adding to herd	Cost, in terms of output foregone, borne entirely by individual.	Benefit enjoyed by all herdsmen – even those free-riding.

Table 2
Forms of Property Regimes

Property Rights System	*Characteristics*
Private Property	Owner enjoys exclusive rights to use the resource and transfer these rights temporarily or permanently.
Open Access	Resources are used by everyone none of whom possess rights which are exclusive or transferable.
State property	Ownership and management is controlled by the state.
Communal property	Use rights are controlled by an identifiable group who devises rules for resource sharing, exclusion, etc.

Figure 1
Costs & Benefits of Establishing Private Property Rights

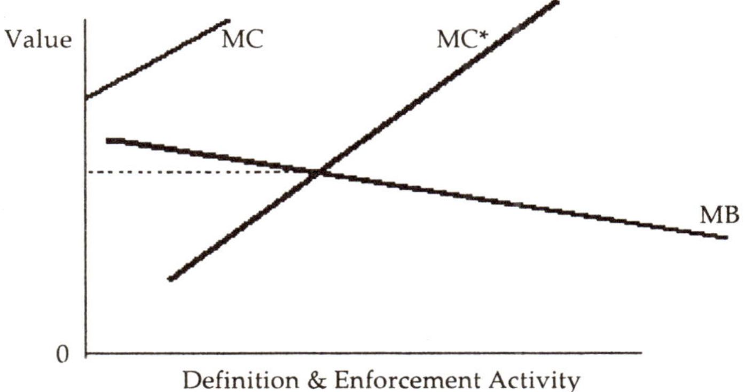

REFERENCES

Alchian, A. and H. Demsetz (1975), 'The Property rights paradigm', *Journal of Economic History*, 33(1), 16-27.

Andersen, Peder (1983), 'On rent of fishing grounds': a translation of Jens Warming's 1911 article, with an introduction' *History of Political Economy*, 15(3), 391-396.

Ardrey, R. (1966), *The Territorial Imperative*, Dell: New York.

Ardrey, R. (1970), *The Social Contract*, Dell: New York.

Baland, J. M. and J. Phillippe (1996), *Halting Degradation of Natural Resources: Is there a role for rural communities*, FAO/Clarendon Press.

Berkes, F. ed. (1989), *Common Property Resources: Ecology and Community-Based Sustainable Development*, Belhaven Press: London.

Bromley, D. (1989), *Economic Interests and Institutions: The Conceptual Foundations of Public Policy*, Basil Blackwell: Oxford.

Burrows, P. (1979), *The Economic Theory of Pollution Control*, Martin Robertson: Oxford.

Cass, R. C. and Julian J. Edney (1978), 'The commons dilemma: a simulation testing the effects of resource visibility and territorial division', *Human Ecology*, 6(4), 371-386.

Cheung, Steven N. S. (1970), 'The structure of a contract and the theory of a non-exclusive resource', *Journal of Law and Economics*, XIII, April, 49-70.

S V. Ciriacy-Wantrup (1971), 'The economics of environmental policy', *Land Economics*, 47(1).

Ciriacy-Wantrup, S. V. and R. C. Bishop (1975), '"Common property" as a concept in natural resources policy', *Natural Resource Journal*, 15, October, 713-727.

Clawson, M. (1974), 'Comment on "Managing the public lands: assignment of property rights and valuation of resources" by A. C. Fisher and J. V. Krutilla', in Haefale, E. T. ed.) *The Governance of Common Property Resources: Papers Presented at a Forum Conducted by Resources For The Future*, Washington D.C. January 21-22, 1974. John Hopkins University Press: Baltimore.

Coase, R. (1960), 'The problem of social cost', *Journal of Law and Economics*, III, 1-44.

Cropper, M. L. (1988), 'A note on the extinction of renewable resources', *Journal of Environmental Economics and Management*, 15, 64-70.

Crowe, B. (1969), 'The tragedy of the commons revisited', *Science*, 166, 1103-1107.

Dahlman, C. J. (1980), *The Open Field System and Beyond*, Cambridge University Press: Cambridge.

de Alessi, L. (1980), 'The economics of property rights: a review of the evidence', *Research in Law and Economics*, Vol. 2, 1-47.

Demsetz, H. (1966), 'Some aspects of property Rights', *Journal of Law and Economics*, 9, 61-70.

Demsetz, H. (1967), 'Towards a theory of property rights', *American Economic Review: Papers & Proceedings*, 57, 347-359.

Edney, Julian J. (1980), 'The commons problem: alternative perspectives', *American Psychologist*, 35(2), February, 131-150.

Furubotn, E. G. and S. Pejovich (1972), 'Property rights and economic theory: a survey of recent literature', *Journal of Economic Literature*. 10, 1137-1162.

Gordon, H. Scott (1954), 'The economic theory of a common property resource: the fishery', *Journal of Political Economy*, 124-142.

Hardin, Garrett (1968), The tragedy of the commons, *Science*, 162, 1243-1248.

Harper-Fender, A. (1981), Discouraging the use of a common resource: The Crees of Saskatchewan, *Journal of Economic History*, XLI, 163-179.

Jodha, N. S. (1986), Common property resources and rural poor in dry regions of India; *Economic and Political Weekly*, XXI [27], 1169-1181.

Jodha, N. S. (1990), Rural common property resources: contributions and crisis, *Economic and Political Weekly*, A65-A78.

Jowett, Benjamin (1967), Aristotle's Politics, (with Introduction, Analysis, and Index by H.W.C. Davis), Clarendon Press: Oxford.

Kramer, R. M. and Marilynn B. Brewer (1984), Effects of group identity on resource use in stimulated commons dilemma, *Journal of Personality and Social Psychology*, 46, 5, 1044-1057.

Larson, B. A. and D. W. Bromley (1990), Property rights, externalities and resource degradation, *Journal of Development Economics*, 33, 235-262.

McKean, M. (1996), Common Property Regimes as a Solution to Scale and Linkage; in Hanna, S, C. Folke and K. Goran-Mäler ed. *Rights to Nature: Ecological, Economical, Political and Cultural Principles of Institutes for the Environment*, Island Press: Washington D.C., 223-243.

Ostrom, E. (1990), *Governing the Commons: The Evolution of Institutions for Collective Action*, Cambridge University Press, Cambridge.

Pasha, Syed Ajmal (1992), 'CPRs and rural poor: a micro level analysis', *Economic and Political Weekly*, 2499-2503.

Posner, R. (1977), *The Economic Analysis of Law*, Boston, Massachusetts, Little, Brown & Co.

Quiggin, J. (1988), 'Private and common property rights in the economics of the environment', *Journal of Economic Issues*, XXII, 4, 1071-1087.

Scott, A.D. (1955), 'Fishing: the objectives of sole ownership, *Journal of Political Economy*; 63, 116-124.

Sen, A. K. (1981), *Poverty and Famines,* Oxford University Press: New Delhi.

Wynne-Edwards, V. C. (1962), *Animal Dispersion in Relation to Social Behaviour,* Hafner: New York.

Wynne-Edwards, V. C. (1965), 'Self-regulating systems in populations of animals, *Science,* 147, 1543-1548.

24

DOES SOCIO-ECONOMIC HETEROGENEITY MATTER IN THE COMMUNITY BASED FOREST RESOURCE MANAGEMENT? AN EMPIRICAL INVESTIGATION FROM ORISSA

Naresh Chandra Sahu and Binayak Rath

INTRODUCTION

Common property resources (CPRs) generally play a very significant role in the rural economy of our country in terms generating income and employment opportunities, and also in improving the food security for the poor and marginal households (Jodha, 1986a). Over the last couple of decades the CPRs, such as forest resources, have been subject to colossal pressure and severe degradation arising out of increasingly growth of population and their growing demands for meeting their livelihood in the rural area (Jodha, 1986b). In view of the significance of CPRs in the rural economy and their increasing degradation in the last few years, our planners and policymakers have adopted a participatory approach in management, protection, and development of forest resources under the aegis of Joint Forest Management (JFM) that was introduced in the National Forest Policy, 1988. Under the new policy, the stakeholders have been ensured property rights in management of forest resources and to receive half of the revenue from harvesting of timbers and non-timber forest products (NTFPs). In response to the national forest policy and to ensure equity and social justice, many state governments have encouraged local communities living nearby forests to form forest protection committees (FPCs[1]) and to participate in forest management. In this regard, Orissa was the first state to introduce such participatory approach in forest management by forming a number of FPCs to revive the degraded state and community forest areas in early 1990s. Till date, 6844 FPCs are registered which are taking care of the management of around 654 thousand hectares of state forests area across the twenty-seven forest divisions of the state. However, there are many issues starting from that of distributional to the participation pertaining to the JFM related activities in the state in particular and developing countries in general, which need to be looked into properly within the framework of institutional economics. In this paper, an attempt has been made to investigate one of the important issues revolving around the JFM management in the state i.e., whether socio economic heterogeneity at household level has any effect on the participation of the individual households in the community based forest resource management.

SELECTED REVIEW OF LITERATURE

The role of heterogeneity in affecting the collective action such ac community based CPR management outcomes are strongly debated as an important issue in the context of natural resource management in both theoretical and empirical literature recently. It is generally argued that heterogeneity in the social, ethnic, economic and political sphere can create impediments to the very performance of the community based resource management because it can make communication and cooperation difficult which directly affect the cost and benefits of users as a whole. It is still a challenge for the social scientists to examine whether heterogeneity affects adversely the collective action such as community based resource management outcomes. There are broadly two schools of thought pertaining to the effect of socio economic heterogeneity on collective action. One school argues that it does while others say, it does not. However, the clear picture relating this case is yet to come up with concrete evidence. Owing to the significance of the illustrated issues in the light of increasingly concerns expressed over the natural resource management, some studies are reviewed focusing both schools of thought provided below relating to the management of the CPRs.

First of all, the issue of the impact of heterogeneity on collective action arose in mid-1960s. It is Olson (1965) who first contended that more inequality may favour the creation of public goods. He explains that inequality (based on endowments, e.g., wealth and land distribution) is good for the management of the common property resources, as these with greater interests may decide to provide the common good even if the poorly endowed group chooses to free ride. Drawing upon recent theoretical advances in the analysis of collective action, Molians (1998) undertook an econometric analysis of the determinants of successful collective action based on a survey of 104 peasant cooperative institutions in Paraguay. His study shows that controlling as much as possible for the specific characteristics of the community and peasant committee, there is a U-shaped relationship between community inequality and cooperative performance. With a view to investigate the impact of heterogeneity on the local level Baland and Platteau (1999) found that local level inequality has an ambiguous impact on the feasibility of the efficient outcome even though better endowed stakeholders contribute more to collective action. In a regulated settings, on the other hand, inequality tends to reduce the acceptability of available regulatory and, therefore, to make collective action more difficult. Dayton and Johnson (2000) developed a simple model of individual household's incentives in a community owned irrigation system and found that social heterogeneity, measured by different villages from which the users of a given system are drawn, is consistently and significantly associated with lower levels of infrastructure maintenance. In a recent study of 18 forests user groups in Nepal Ostrom and Varughese (2001) found that heterogeneity (location, socio-cultural and wealth disparity) is not a strong predicator of the community resource management outcomes (collective action) using Logit regression model. Similarly, Somanath *et al.* (2002) examines the effects of village level of heterogeneity in caste and land ownership, and of female membership on the community based forest management in the Indian Central Himalayas. The study found no evidence that caste heterogeneity or female membership of the van panchayat have any effect by using the logit regression model. But it is found that greater equality in land ownership may enhance community action.

On the contrary to the findings of the above reviewed studies, some scholars argue that heterogeneity hampers community resource management outcomes. There are at least three different arguments put forward by scholars why heterogeneity may be inimical to the success of collective action such as community based resource management outcomes. First, according to Alesina , Baqir, and Easterly (1999) suggest that tastes for public goods may differ across groups and this can reduce the willingness to pay for public goods which may not be the kind of that one's own group wants. Second, very often-political scientists argue that social divisions may throw up politicians who seek to provide private benefits to particular groups rather than providing public goods. Third argument states that heterogeneity makes it harder for communities to reach an agreement about the sharing of the benefits or costs of collective actions. Inequality in power, for instance, may mean that equal division would be unacceptable to the powerful, while any other distribution may be subject to conflict. The basis of this argument is that heterogeneity removes a natural focal point for the consensus and, simultaneously makes groups uncertain about other group's preferences, thus making agreement less likely as each group tries to drive a hard bargain, one that may be unacceptable to other groups. This argument calls for the examination of heterogeneity in the context of community resource management. For instance, let's take the asset heterogeneity for the efficient common property resources managed outcomes here for explanation. As it is know known that the common property institutions are self organized institutions, and set rules pertaining to the access, use and harvesting and conservation of common property resources such as forest, confined to well defined and delineated boundary within a particular area. It is obvious fact that landless and poor households would like to collect fuel woods, food items fodder, etc., frequently from the common property resources by going against the set rules of limited access in order to meet their daily needs for substance of their livestock and livelihood why because they do not have other private resources to collect those mentioned stable requirements whereas the large and rich households easily meet their daily needs of fuel woods, fodder, etc., since they hold enough private assets such as land and other supplementary private property to support their needs without defying the use and access rules . The restricted access rules affect adversely the benefits of the poor households from commons while these don't have any significant effect. Thus, the heterogeneity of asset would certainly have negative impact on the probability of the efficient community based managed resources outcomes by adversely affecting interest of some group of households who are member of the said institutions. However, the topic is still a challenging subject among the scholars to unravel the relation between heterogeneity in its all forms and collective action over a couple of years now in the context of natural resource management.

STUDY AREA AND METHODOLOGY

The study was carried out in four forest divisions of Orissa where community based forest resources management institutions under the aegis of JFM have been implemented since early 1990s. For this study, twelve FPCs from four forest divisions are selected, of which six FPCs (Kendumundi,Kandhakhal , Bhezapadar, Hatikut, Budhikhamari and Khasadiha) are located in the tribal dominated districts of Kalahandi and Mayurbhanj and rest six FPCs (Sudhasahi, Baniasahi, Madanmohanpatna, Kushapplah, Bimpada and Kotpallah) are situated in the non-tribal districts of Angul and Khurda. The land uses pattern of the study villages are categorized as cultivated land, area not available for cultivation, village forest area, and culturable waste and grazing land. Apart from these classification these threes villages have provided access rights to the protection and management of the state forests. Though the caste structure is heterogeneous and comprising of various castes, such as, Brahmins, Khandayats, Sudras and Tribals, but the tribal constitute the dominant grouping the two tribal districts whereas the non-tribal form major caste in other two non tribal districts. Moreover, it has been found that the major percentages of the total population in these villages are subsistence farmers and forest dwellers. The immediate source of livelihood of the study area is the collection and selling of NTFPs, such as fuel wood, bamboo, kendu leaves, *Sal* leaves, lakh, honey, *gum, lak, mahua* flower, medicianal herbs, etc.

Map Showing the Location of the Study Area

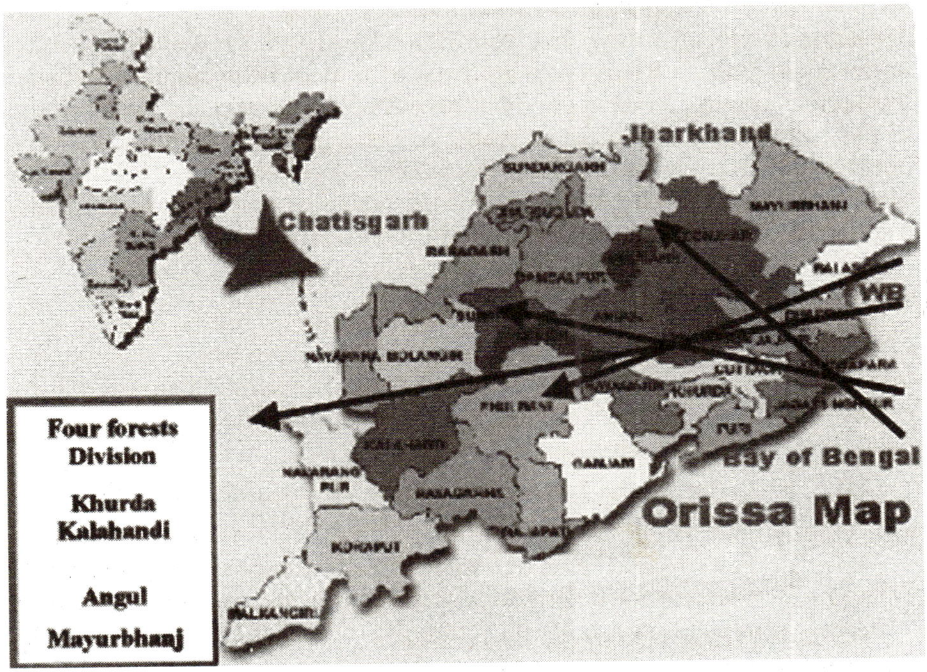
Jharkhand
Chatisgarh
WB
Bay of Bengal
Orissa Map
Four forests Division
Khurda
Kalahandi
Angul
Mayurbhanj

The data collection of the entire study comprises of two phases: secondary data collection and primary data collection. Under secondary data collection, the information about total FPCs villages and the history of their establishment, their land use pattern, demographic trend, and access to forests resources were collected from their respective forest blocks. Primary data on household level variables, their dependency on forest and management of community forestry was randomly collected through a semi close ended household questionnaire for a period of two months from January to March 2005, in twelve villages. The main criteria for the categorization of the households are: land owned, number of livestock ownership and income from off-farm activities. The household is classified into three major categories namely, medium farmer, small and marginal farmers and landless poor agricultural labourers. A total of twenty per cent households are randomly selected from each FPCs. In the field investigation 318 households are covered by using semi-closed questionnaires covering three important areas: demographic, landholding and off farm production activities and their participation in the natural resource management and utilization.

REGRESSION MODEL AND RESULTS

The econometric analysis to find out the impacts of socio economic heterogeneity on the formation of collective action at village level could not have been carried out due to the paucity of the observations. At village level we have collected the information about only 12 villages, which are not enough for adopting any regression analysis. However, we have undertaken Logit's Regression analysis to find out the relation between the socio economic heterogeneity at household level and the participation in the community based forest resource management activities. To carry out this exercise, we have used collected cross-section data of households. Moreover, the following household level variables and functional form of the regression equation are generated for this purpose:

HPCFI = f (HC, HS, HL, HLO, HTO, HI, HAW)

HPCFI = A household's decision on whether or not to participate in FPC related activities (has a yes/no value). The payment of membership fees and attending the meetings of the FPCs by households are used as the proxy of participation variable.(Dummy variable, = 1 if houdols participates in FPCs activities, 0 = otherwise)

HC = Caste of the household (Dummy variable, = 1 households belongs to upper caste, 0 = otherwise)

HL = Educational level of household head has been used (Dummy variable =1 if the household head is literate, 0 = otherwise)

HS = family size of the household

HLO = Size of land holding of household

HTO = Number of trees owned by household

HI = Total household income

HAW = Total household animal wealth.

Dummy variable are used for two variables: caste and educational level of the households in the regression model. The Logit Regression model is used to analyse the categorical variables HPCFI pertaining to household's decisions to participate in FPC institution. This methodology facilitates the eliciting of effects of several predictor variables, which may be quantitative, categorical or mixture of the two, on dichotomous or binary response variable which depicts decisions like participation or non participation in any collective action or basically yes /no kinds of decisions on the parts of the respondents.

The functional form of Logit model for our present analysis is as follows:

The basic for m of multivariate logistics function with Z as the predicator variable, which itself is a combination of several predicator variables ($X_{k's}$) as

$$Z = b_0 + b_1 X_1 + b_2 X_2 + \ldots\ldots\ldots\ldots + B_k X_k \quad (1)$$

and thus the functional form of the model becomes

$$P_k = E(Y = 1/X_k) = 1/(1 + e^{-(b_0 + b_2X_k)}) \quad (2)$$

$$P_k = 1/1 + e^{-z} = e^z/(1 + e^z) \quad (3)$$

Equation (3) represents the logistic distribution function.

If P_k, the probability of participating in FPC is given by equation (3), then $(1 - P_k)$, the probability of now participating in VSS activities, is

$$1 - P_k = 1/1 + e^z \quad (4)$$

Therefore we can write

$$P_k/1 - P_k = (1 + e^z)/(1 + e^{-z}) = e^z \quad (5)$$

Now $P_k/(1 - P_k)$ is simply the odd ratio in favour of participating in FPC activities by a household. Now if we take the natural log of equation (5), we get the following interesting result,

$$L_k = \ln(P_k/1 - P_k) = Z \quad (6)$$
$$= b_0 + b_2X_k$$

L is the log of odd ratio, not only linear in X but also linear in the parameters. Therefore L is called the Logit and hence the equation (6) is called Logit model.

In which P is the probability of participation in FPC activities in the present discussion and thus W depicts odds of participation. In fact, the reference category in the present model is non-participation in FPC and P thus provides the probability of participation in VSS.

In this model Maximum Likelihood (ML) estimation procedure is adopted for obtaining the parametric estimates of the structural coefficients (b_{js}) of the model. Gujurati (2003: 633) provides mathematical details of the ML estimation procedure. The estimated structural coefficients (B_{js}) by the ML estimation procedure and their mean values are in turn utilized to draw the estimates of probabilities of participation or non-participation in the FPC.

The behavioral decision of the household is considered here that of whether or not to participate in the different activities that lead to creation of well-specified rights in common property resources. This kind of decision strengthens the institutions created with outside intervention and is measure of the extent to which institutional innovation has been internalized. In this study we have identified the conditional variable in terms of household's monetary contribution/membership fee or labour contribution for the plantation, protection and development of forests as dummy of participation in the VSS activities. If a household is paying monthly membership fee/labour contribution for the protection and development community forests then this can be considered as the effective participation by the concerned households, otherwise the household will be considered as no participant in the FPCs activities.

The estimated coefficients of the Logit Regression Model for participation in FPCs activities along with the estimated structural coefficients and their standard errors are presented in the following table.

Table1
Effect on participation in VSS activities: Results of the Binary Logit Regression Analysis

	HPCFI			
Predicator Variable	*Coefficient*	*Standard Errors*	*Z Statistics*	*P > |z|*
HC	–0.7292833	.422277	–1.73	0.084
HL	1.568219	.5798305	2.70*	0.007
HS	0.1538104	.1046845	1.47	0.142
HLO	–0.2281549	.0744253	–3.07*	0.002
HTO	–.0074962	.0053884	–1.39	0.164
HI	–.0000131	5.46e–06	–2.40*	0.016
HAW	.1073311	.0973566	1.10	0.270
Constant	3.750914	.7190399	5.22	0.000

LR $chi^2(7df) = 77.88$
Prob > $chi^2 = 0.0000$
Log likelihood = –94.67309 Pseudo $R^2 = 0.2914$
Total number of observation 318

Note: *significant

The perusal of the above table reveals that educational level of household (HL), land owned by household (HLO), and household income have turned out to be highly significant effect on the probability of participating in the community based forest resources management under JFM programme. All the three variables have expected signs. The educational level of the households has a positive impact on the probability of participating in the community based forest management. In other words, the higher is the educational level of household the larger is the possibility of participating in the formation of collective action. In fact, it states that educated people are more conscious abut the conservation of forest resources and their crucial role in maintaining a sustainable environmental resources in the rural area. However, the negative sign of the two variables HI and HLO clearly establishes that the higher the income and private land of the households the lesser is the probability on the part of the household to participate in the collective action such as FPC activities. The reason behind the less probability on the part of the large landholders in joining the FPC initiatives is that the rich farms depend less on the forest resources to meet their livelihood requirement in terms of fuel wood, fodder and NTFPs. They have private orchards and gardens to cater their fuel wood and fodder needs. Therefore, rich and medium farmers are less dependent on the forest resources unlike marginal, small and poor farmers in the study area. Though other variables, such as caste of the household, family size of the household, number of tree owned by household and total animal wealth of the household have expected signs but they don't have significant impact on the independent variable i.e., participation in FPC activities.

CONCLUSION

The estimated results of our study entail that the socio-economic heterogeneity has an ambiguous impact on the probability of the participation and formation of community based forest resources management in the state. The interesting findings of the study are that cast heterogeneity has not significant relation with the collective or FPC activities contrary to the views of some scholars. However, the educated level of households has significant effect on the probability of participating in the community based forest resources management. In addition, the main finding of the study is that the land and income inequality may hinder the collective action. Though the results of the study lack any clear-cut conclusions about the role of heterogeneity on the collective action but it has significant policy implication in the management and formation of FPCs under the flag of JFM programme. The state and central government should take not of the crucial role of education in the formation of collective action of community based forest resources management. Before encourgaging the formation and participation of households in the JFM forest department and concerned official should promote education to make this endeavor successful in achieving forest resource conservation along with improving the livelihood patter of the stakeholders in the forested area. The role of socio economic heterogeneity should be looked at thoroughly in order to accomplish the success of collective such as community based forest resources management in a socially and economically diversified country like India.

NOTE

1. FPC is locally known as Vanna Surakhya Sameeti (VSS) in Orissa.

REFERENCES

Alessina, Alberto, Reza Baqir, and William Easterly (1999), Public goods and Ethnic Divisions, *Quarterly Journal of Economics*, 1149(4): 1243-84.

Baland, J. and J. P. Platteau (1999), The Ambiguous Impact of Inequality on Local Resource Management, *World Development*, 27(5): 773-788.

Dayton-Johnson, J. (2000), Determinants of Collective Action on the Local Commons: A Model with Evidence from Mexico, *Journal of Development Economics*, 62: 181-208.

Gujurati, Damodar N. (2003), *Basic econometrics*, 4th ed., McGraw-Hill, Singapore.

Jodha, N. S. (1986a), Common Property Resources and the Rural Poor in Dry Regions of India, *Economic and Political Weekly*, 21(27): 247-264.

Jodha, N. S. (1986b), Population Growth and the Decline of Common Property Resources in India, *Population and Development Review*, 11(2): 169-181.

Molians , J. R. (1998), The impact of Inequality, Gender, external Assistance, and Social Capital on Local Level Collective Action, *World Development*, 26(3): 413-420.

Olson, M. (1965), *The Logic of Collective Action: Public Goods and the Theory of Groups*, Cambridge: Harvard University Press.

Ostrom, E. and Varughese G. (2001), The Contested Role of Heterogeneity in Collective Action: Some Evidence from Community Forestry in Nepal, *World Development*, 29(5): 747-765.

Somnath, E. *et al.* (2002), Collective Action for Forest Conservation: Does Heterogeneity Matter?", *http://www.santafe.edu/files/*

25

UNCERTAIN LIVELIHOODS: SURVIVAL STRATEGIES IN *CHAR*LAND ENVIRONMENTS IN INDIA

Kuntala Lahiri-Dutt and Gopa Samanta

INTRODUCTION

In developing countries increasingly large numbers of people are living in marginal environments characterized by poor living conditions and extreme resource constraints. *Char*lands are temporary, low river islands common to deltaic regions, and epitomize these characteristics. *Char* inhabitants are the poorest and most disadvantaged; yet have adapted to this environment and have developed livelihood strategies to ensure their survival. This paper focuses on livelihood strategies of *char*land inhabitants in the highly dynamic *char*lands of the Damodar River in Southern Bengal, India. Our study owes a great debt to Baqee's research (1998) on *char*lands in Bangladesh describing them as *'Allah jaane'* (God knows, or whatever will be, will be) lands and focusing on the process of occupancy, dislocation and resettlement. Other research on *char*lands (such as that by Chowdhury, 2001) examines issues such as disaster mitigation, but so far little has been written about livelihood strategies in these temporary lands.

All *char*land dwellers have well-established livelihood strategies characterized by highly diverse production and income-generating activities. These include farming, livestock rearing, fishing, petty trading and working as itinerant wage labourers. However, for the extreme poor, these strategies merely permit survival rather than the accumulation of sufficient assets to overcome poverty (Elahi *et al.*, 1991; ISPAN, 1995).

Examining these livelihood strategies is important to understanding the ways people maximum utilization of scarce resources for survival, and cope with the vulnerabilities brought about by natural hazards. It has been noted (Valdivia and Gilles, 2001) that water, land, livestock, crops and knowledge all play significant roles in the livelihoods of most of the world's rural households. In *chars*, the vulnerability created by flood and riverbank erosion plays a dominant role in the livelihoods of inhabitants who are largely excluded from mainland services and infrastructure. These people do not constitute a priority for local or state governments, as the legal status of their land is not easily determined, though most inhabitants are illegal immigrants occupying public land.

RESEARCH METHODOLOGY

Participatory research methods were used to explore the livelihood strategies of poor migrant people living in *char* Gaitanpur as this offers a creative approach to information sharing. The participatory approach is based on field visualization, interviewing and group discussions promoting interactive learning; shared knowledge and flexible yet structured analysis (IDS, 1997).

Both primary and secondary sources of information were collected for this study. Secondary information is limited to the 2001 census figures for *char* Gaitanpur. The primary data collection was undertaken between June 2002 and September 2004 and used both quantitative and qualitative methodologies. A structured questionnaire was used to survey each of the 202 households in *char* Gaitanpur and this work was carried out with the help of eight trained assistants. Fifty women of different age groups and economic classes were also surveyed to explore the gender roles in *char* livelihood strategies. Another set of qualitative survey data was collected from 50 women and 50 men through personal interviews and recording of oral histories, and informal small group discussions took place on aspects of different livelihood strategies.

THEORETICAL BACKGROUND

Various scholars have defined livelihood in different ways. The first connotation of the term is of course economic well-being; Ginguld *et al.*, (1997: 577) see livelihood as a set of strategies undertaken by a particular household in order to secure its economic well-being and especially its long-term survival. This draws from a tradition set by Pearson in 1977, who analyzed the emergence of economic transactions and the origin and development of trade, money and market as part of human livelihood. Some recent studies such as that by Valdivia and Gilles (2001: 7) also see livelihood strategies as a portfolio of activities and the social relations by which families secure or improve their well-being or cope with crises. According to Frankenberger (1996) livelihoods consist of on-farm and off-farm activities which together provide a variety of procurement strategies for food and cash.

These definitions tend to emphasize the economic dimension of living, but others have put more importance on the overall processes of securing a survival. For example, according to Blaikie *et al.* (1994: 9), the word means the command an individual, family or other social group has over an income or bundles of resources that can be used or exchanged to satisfy its needs. This may involve information, cultural knowledge, social networks, legal rights as well as tools, land or other physical resources. Other experts too have chosen to see it as a broad amalgamation of various elements. Ellis (2000: 10) defined livelihoods as 'the assets (natural, physical, human, financial and social capital), the activities (strategies of use), and the access to these (mediated by institution and social relations) that together determine the living gained by the individual or household'.

The livelihood approach to understanding the survival strategies of poor people, as well as development processes, has become increasingly popular in the last decade. This has been a direct response to making the planning process work for the poor according to what is important for them. Overhauled *et al.* (2004) considered the study done by Chambers in 1983 as the pioneer work on livelihoods of the poor. In 1989, Chambers noted that livelihoods are more sustainable when households have secure ownership of, or access to, resources and income-earning activities, including reserves and assets, enabling them to offset risks, ease shocks and meet contingencies. Since then a number of studies have been carried out on different aspects of livelihood (see Ashley *et al.*, 2000; Bebbington, 1999; Brock and Coulibaly, 1999; Bryceson, 2000; Bryceson *et al.*, 2003; Chambers and Conway, 1992; Francis, 1999; Marzano, 2002; Scoones, 1998; Yaro, 2002). Developmental agencies and governments are increasingly using the concept of livelihoods in the design of policies, projects and programmes.

Since the late 1980s, a new angle in these literatures has been the emphasis on sustainability. Consequently, many livelihood studies carried out by development agencies to alleviate poverty have focused on examining sustainability as an insurance against total breakdown of livelihoods (Ashley and Carney, 1999; Brock, 1999; Carswell *et al.*, 2000; de Haan 2000a; DFID, 2002; Toufique, 2001). The concept of sustainable livelihoods used in these studies relates to the conditions of poverty, well-being, capabilities, resilience and the natural resource base of households and communities.

A number of studies in livelihood literature focus upon the role migration can play in the livelihoods of poor rural households (de Haan, 1998, 1999, 2000b; de Haan *et al.*, 2000; Mc Dowell and de Haan, 1997). It is now generally recognized that migration is part of the normal livelihood strategy of the poor (see Hossain *et al.*, 2003 and Siddique, 2003 for work on Bangladesh). However, the rate of migration, both permanent and temporary, for improved livelihood opportunities increases at times of socio-economic distress, political crisis and/or natural disasters.

Another set of literature has looked at the issue of diversification of livelihood strategies as the critical element in creating better options for sustainability (Deb *et al.*, 2002; Ellis, 1998, 1999; Toulmin *et al.*, 2000), and others have assessed how people can develop strategies to diversify their survival base in order to better cope with change (Ginguld *et al.*, 1997; Karnath and Ramaswamy, 2004).

Women in poor communities may constitute the poorest or weakest group, bearing the burden of care for the children and family. Thus, the ways women and men seek and sustain a livelihood are often different. The following studies (for example Cleaver, 1998; Francis, 1998 and 2002; Hapke and Ayyankeril, 2004; Masika and Joekes, 1996; Valdivia and Gilles, 2001) look at livelihood as a gendered activity. A few studies such as the one by Francis (1998) analyze the bargaining power of women and their access to household resources along with their contribution towards the livelihoods of the households. She showed that women struggle to earn a livelihood for the family and to run the household, but do not enjoy significant additional power as a consequence (Francis, 1998: 75).

CHAR GAITANPUR

Char Gaitanpur is located on the Damodar riverbed in Burdwan district, West Bengal and comprises a total area of 2.5 sq. km. Burdwan town, the district headquarters of about 300,000 people, is located 3 km to the north and provides an important labour and produce market for *char* inhabitants. The position of *char* Gaitanpur has changed several times due to the frequent shifting of the thalweg (main flow channel) of the river. Bhattacharyya's (1998) study of the Damodar refers to Dicken's map of

1854 showing Gaitanpur as a marginal bar attached to the south bank of the river. In September 1959 there was a major flood and the Survey of India map of 1970 showed it as a mid-channel bar (island *char*). During the devastating flood of 1978 the course of the river completely changed and *char* Gaitanpur again became attached to the north bank of the river. The old course of the river in the north has become an agricultural field, though is not registered in the government settlement records, while on the southern bank, some land (which no longer exists) is still registered in these records.

The physical character of the *char* Gaitanpur is somewhat different from other *chars* located in the active delta region of the Ganga–Padma and other rivers in deltaic Bengal. Regular flooding, an important characteristic of these *chars*, is absent here. The problems of annual inundations, shifting land and the consequent regular conflict that can accompany the re-demarcation of land that has been covered by sand and alluvium, are less intense in this *char*. Gaitanpur provides a comparatively secure environment in which migrants can settle, while at the same time presenting a marginal environment in which established local communities may not wish to live. Property boundaries have become more or less permanent, which offer residents a slightly safer, but not adequate, livelihood. River erosion is a greater threat to the dwellers of Gaitanpur *char* than floods. The film of loamy soil above the substantial layer of sand is only about 3-6 inches thick and therefore highly susceptible to erosion. Bank erosion reached its peak during the 1978 flood when half the *char* was lost to erosion. However, this flood also deposited sand and alluvium on the old course of the river channel and made an extensive area available for cultivation. Some households have been displaced by erosion up to three times during their time in this *char*.

Peopling Process

Char Gaitanpur was first settled in about 1947 by a group of Bihari boatmen[1]. As land was cleared and prepared for farming more landless people moved to the *chars*. This process of peopling continued until around 1954, by which time the initial occupancy of the land in *char* Gaitanpur was complete. Between 1954-55 the official land survey commenced and the Bihari settlers received their legal rights to the *char*lands. At present there are 58 Bihari families living in this *char*. It appears that those who have remained in the *char*land have done so due to the complete absence of alternative subsistence options. This lack of choice is echoed in the voices of Bangladeshi immigrants who began migrating to *char* Gaitanpur in 1988. Since 1990 many Bangladeshis have come to live in these *char*lands - about 95 per cent from Khulna district. Through the development of a network of kin, many of those who migrated to other parts of West Bengal have now come to live in the *chars*.

Population Composition

Char Gaitanpur is settled by two distinct groups of migrants – Biharis and Bangladeshis – but the population composition has changed rapidly since the 1990s. Biharis were the majority until 1990, however they are now a minority group, occupying approximately one third of the total houses (58 out of 199). This situation has arisen because of their failure to develop agricultural skills, being fishing people traditionally, and much of their cultivable land has been sold to Bangladeshis. Out-migration of young Biharis continues as the younger generation has no interest in farming and prefers informal jobs with fixed wages. As a consequence they leave the *chars* as soon as they find a job on the mainland. Migrants also use *char*lands as places of short stay on the way to more stable locations. The ease of entry into *char*land society, due to its relative physical isolation from mainstream life, attracts new migrants, especially those crossing the border without papers. However, many try to leave for Burdwan town as soon as they are economically secure. The total population is increasing at a spectacular rate (11 per cent per annum compared with the Indian national population growth rate of 2 per cent per annum in the last decade). In 2001 the population of *char* Gaitanpur was 797, and had increased to 1,068 by 2004.

Settlement Pattern

About one-third of *char* Gaitanpur is higher ground and houses are built on this ground, on either side of a wide dirt track that cuts through the *char*. The marginal lower lands are used for cultivation, and in most cases the land adjacent to a house is 'owned' by the householder; this ownership arrangement is accepted, understood and respected by all inhabitants. This proximity of land has facilitated the employment of family labour, especially women, in the field and allows for continuous supervision of the vegetables they produce. The Bihari neighbourhood is separated from the Bangladeshis, which continues to grow in size.

Livelihood Assets

Livelihood assets are a combination of natural, physical, social, human and financial capital upon which a group or community develop their livelihood. Swift (1989) defined livelihood assets as a wide range of tangible and intangible stores of value or claims to assistance, which can be mobilized in a crisis. The sustainability of livelihoods of any community largely depends on their access to assets, which they can exchange or cash-in during times of crisis (Yaro, 2002). In 1989, Berry noted that access depends on participation in a variety of social institutions, as well as on material wealth and market transactions. In *char*lands of Damodar, the main livelihood assets take the form of physical capital including cultivable lands for cropping, pasture for livestock rearing, river water for fishing and washing-cleaning, subsurface water for drinking, and riverbed sands for quarrying. Natural capital includes the fertile silt deposited by floodwaters, and the readily available groundwater that facilitates irrigation. As the river has dried up, attention has focused on land-based assets. Social capital is represented by kin relations and the community support provided by people living in this vulnerable environment. The human capital of these people includes their farming skills and their physical labour. All these assets are used by the *char* residents to generate financial capital, which rarely surpasses the livelihood needs of households to create sufficient surplus for further investment.

Livelihoods in the Early Stages of Settling

> 'I can still remember the hunger pains of those days. I came here at the age of 16 just after my marriage. Sometimes I could not tolerate the long gap between the only two meals of the day with which I was not accustomed in my parental home. However I tried to cope with those situations without retaliation, because speaking about pain does not reduce it rather gets enhances.
>
> Sumaria Rajbhar

> We spent day after day by taking only flour prepared from seeds of *mesta*[2] jute which was the most common crop of the *char*. Getting rice or bread for either lunch or dinner was really a rare experience of life in those days.
>
> Saraju Chowdhury

The early Bihari settlers lived by catching fish, rearing cattle or labouring in nearby mainland areas. The new settlers did not plant crops because they lacked farming skills and a knowledge of irrigation. *Mesta*, wheat, maize and pulses were the main rainfed crops produced without effort on the limited agricultural land. Acute poverty accompanied people then and now, but the cultural landscape of the *char*lands has changed with the immigration of Bangladeshis who have an intimate knowledge of low-lying lands and their ecology. It is the Bangladeshi immigrants who have given rise to the present livelihood strategies of the *chars*.

PRESENT LIVELIHOOD STRATEGIES

Agriculture

As in many places in South Asia (see DFID, 2002; World Development Report, 2000), agriculture forms the mainstay of the economy in the *char*lands of Damodar (see Table 1). About 73 per cent of households in *char* Gaitanpur have some farming land, although most are best described as either small-scale or marginal farmers. *Char* agriculture is characterized by the small size of land holdings (Table 2) and labour intensive cropping methods. Double cropping is practiced all over the *char*land, but multiple cropping is limited to lands owned by farmers with shallow pumps for irrigation (26% of the farming households). Significant crops include vegetables, which are sold in the local markets, and paddy, which is a household staple.

Table 1
Household Sources of Livelihood

Sources of livelihood	*Number of households*	*Pent cent of households*
Cultivation	29	15
Cultivation & agricultural labour	87	43
Cultivation & petty trading	23	12
Cultivation & service	5	2.5
Cultivation & others	28	14
Agricultural labour	24	12
Agricultural labour & petty trading	3	1.5
Total	199	100

The dry season (winter and early summer) is the main cropping season of the *char*. During the *kharif* or monsoon main cropping season (July to October), poorer farmers keep their lands fallow as they cannot arrange their own minor irrigation system during this time of high evaporation. Farmers with more than three acres of land are defined locally as 'big farmers' who usually have their own minor irrigation system and can hire additional labour for farm work. Small-scale farmers (owning 1-2 acres and less than 1 acres of land respectively) sometimes have shallow pumps but generally cannot afford to hire labour, and rely on family members to work on the farm. Farmers receive a better price if they take their vegetables to the nearby Burdwan wholesale market.

Table 2
Proportion of Agricultural Land under Different Holding Sizes

Size of land holdings (acres)	*No. of households*	*Pent cent of households*
Less than 1	76	52
1 – 2	59	40
3 – 4	11	8
Total	146	100

Risks in Agriculture

Farmers usually face *two* major risks: floods and price fluctuations. Floods are a natural hazard that always accompanies the life of *char*land people. Most of the lowland farmers usually keep their land fallow in the rainy season to cope with this risk. Pest attack sometimes affects the production of vegetables and reduces the yield. Fluctuating crop prices are the most significant risk to farmers. Potatoes are the staple crop in *char*lands as they grow well in sandy soils and being a winter crop the risk of flooding is low. Over the last ten years the average price of potatoes has varied between Rs. 120 and Rs. 300 per 100 kg (Table 3). Profit and loss has varied accordingly. *Char*land farmers cannot hold their produce in order to receive a better price as they do not have access to cold storage, and the costs associated with transporting potatoes to cold storage are also high. Most farmers produce potatoes with loans from Mahajan at a high interest rate, and they need immediate cash to repay the loan, even if they have to bear a loss. All farmers grow potatoes despite the high investment cost and high risk of price fluctuations, as it provides a very high return if the price does not fall. Ganesh Halder described the importance of potato farming very clearly: 'If we get a good price then the return from potato farming on only one acre is sufficient for the annual cash expenditure for the poor people like us.'

Table 3
Yields, Prices and Profits Associated with Growing Potatoes

Average yield per acre (kg)	*Price per 100 kg.*	*Total Price (Rs.)*	*Total expenditure per acre (Rs.)*	*Profit (Rs.)*	*Loss (Rs.)*
15,000	300	45,000	21,000	24,000	–
15,000	150	22,500	21,000	1,500	–
15,000	120	18,000	21,000	–	3,000

Wage Labour

The livelihood of the landless is characterized by the seasonality of activities (Makita, 2003). Landless people usually earn their livelihood as wage labourers in agricultural fields or in sand quarries in the riverbed. Younger people prefer to work in the sand quarries because they receive slightly higher wages. Agricultural labourers can only earn up to Rs. 50 (~US$ 1) per day while sand quarrying is done on a piece-rate contract basis by a group of four to five people, and can earn between Rs. 80 to 120 (~US$ 1.8 – 2.4) depending on the volume of work. Many older people are unable to work in the sand quarries because of the heavy physical labour requirement and must rely on seasonal agricultural work.

Rearing Livestock

Livestock rearing – particularly goats, ox-bullocks, cows and pigs (Biharis prefer to raise pigs) – is an important subsidiary livelihood activity. Fallow land is usually used as pasture for domestic cattle and agricultural land is used for grazing during the lean cropping season i.e. summer and the rainy season. The cost of rearing cattle is very low during the lean season as the animals only need to be fed at night, however costs increase in the peak season (October to March) when agricultural land goes under crop. During this time people tend to graze their cattle along the roadsides or in small fallow patches. Fodder usually has to be bought for dairy cattle, while goats are managed by collecting tree leaves. Older people, especially women, usually look after the animals to supplement their family income, however women rarely have a say over the profit made. Women prefer to rear milk cows as domestic cattle because of the multiple contribution they make to their daily livelihood. Fulmati Mondal, one of our women participants, made it clear: 'With goat, I can earn some cash from selling it after a certain period of time. If I rear ox or bullock then they are used for ploughing the land, which ultimately help men in their field jobs. However, the cows give us milk, which lessens our burden of providing food for children to a large extent because with milk I do not need anything else with rice. I can manage the family's daily meals with rice, some mashed potato and milk. This reduces both the cost of food and the time for cooking as well as improving the health of my kids.'

Petty Trading

Petty trading is an important livelihood activity for some families and helps to earn an average between Rs. 40 to 50 per day depending on the market price, which is comparable with the wages for agricultural labour. Some households depend entirely on petty trading, while for others it provides a supplementary income. In *char* Gaitanpur 53 people (20 males and 33 females) from 199 households are engaged in petty trading. Most sell vegetables from door to door in nearby Burdwan town, and only two are engaged in fish trading. For various practical reasons farmers prefer not to sell their vegetables directly to petty traders. We quote Sachin Mondal on this point: 'We prefer to sell our vegetables to the wholesaler rather than to the petty traders. Petty traders usually take smaller quantities of different vegetables, which is not feasible for us to sell. We sell our vegetables in large quantities at a time and we don't have the time to weigh the vegetables in smaller quantities. Again sometimes it becomes difficult to get the cash back from the petty traders. Rather we get cash for our products on a daily basis from the wholesale traders.' This increases the economic circuit and the number of jobs involved.

The age of men involved in petty trading is usually between 25 to 40, whereas for women it is 35-55. Younger women are not usually allowed by their partners to get involved in petty trading. The responsibilities of childcare also mean they are unable to leave their homes for prolonged periods. Many men choose petty trading because it provides a source of income throughout the year. Women usually choose this occupation to get access to the outside world and to improve their economic and social status within the household.

Fishing

The availability of fish has reduced remarkably with the reduction in river flows and rising levels of pollution in river water. Consequently, fishing has lost its significance as a major source of income for people living in *char*lands. However people still fish for household consumption or for subsidiary income. Their fishing activities are influenced by the availability of fish and the ease of fish catch during different seasons. The catch varies from 250 gm to 10 kg; increasing in summer and decreasing in winter. The price paid for fish varies from between Rs 40 and 100 depending on the type and size of fish.

Diversification of Livelihoods

Diversification of income and assets is closely linked with livelihood strategy (Datta and Hossain, 2003) as it offers people options for coping with crises. Household incomes are usually generated from different and varying sources depending on the income-earning opportunities open to each individual or household (Yaro, 2002). Diversification across income sources helps households to combat instability in income and thereby increases the probability of their maintaining livelihood security (Deb *et al.*, 2002). While agriculture remains the main source of income for the majority of the households a degree of diversification in livelihood strategies has taken place over the last ten years. Proximity to Burdwan town and the development of transport infrastructure has facilitated this diversification.

Diversified livelihood strategies are more common among poorer people than among those who are relatively better off. Another study carried out by Deb *et al.* (2002) in two villages of Andhra Pradesh, India observed that 'whilst there are a small number of cases where diversification has enabled households to lift themselves significantly above the poverty line, the overwhelming experience of diversification is as a coping strategy' (p. vi). In our study area, diversification is taken up by landless households and small, marginal farmers as a common livelihood strategy. Large farmers occupying better land on higher ground usually produce three vegetable crops per year using their own minor irrigation systems. Members of these farming households are occupied with production and marketing activities throughout the year and do not have the scope or need for diversification of livelihood activities.

Livelihoods During the Lean Season

The lean agricultural season i.e. April to September is a hard time for most *char*land people, especially landless agricultural labourers. The relatively better off farmers can support themselves during this season with the profit they make from winter crops. Farmers, who own minor irrigation systems can produce summer crops, especially rice, although on limited stretches of land. The rest keep their land fallow and try to find other forms of employment and income. As most of the land remains fallow between April and September, some agricultural workers go to the mainland to look for seasonal work, but women cannot travel such long distances, as they have to look after the household.

An important livelihood strategy during the lean season is the farming of vegetables on a subsistence basis in the courtyard of their houses. Ashalata, a woman agricultural labourer stated: 'we grow vegetables in our courtyard to meet the daily needs of vegetables. We have four hens, which give us egg sometimes. We sometimes go to catch fish from the river for our own consumption. We do not have acute poverty and crisis for food neither we have enough surplus. However, we face problem if any unseen expenditure comes like any of our family member falls sick. Then we have to take loan from the *mahajan*[3].'

Gender Roles in Livelihood Activities

*Char*land women make a significant contribution to managing the household. Gender role differentiation is very strong and women's work is identified with the private sphere of the household, and all domestic work is considered to be the responsibility of the women. In addition to their household chores, women are burdened with heavy manual jobs that provide an uncertain income. In terms of their decision-making power, women are marginal both within the household and within society. With little or no security made available through the state support system, women are forced to devise innovative ways of making a living in hostile circumstances.

Our survey revealed a persistent social stigma attached to women's participation in the workforce. Women's employment continues to be considered a sign of poverty, therefore only poorer women, constituting 28 per cent of total women surveyed, earn a livelihood outside their home. Younger women in poorer households often take their children to their place of work. Working women who have other female members of the household (mother-in-law, daughter or daughter-in-law) to share the domestic burden consider themselves lucky. Thus, despite the positive impact of women's labour force participation on human resource allocation and economic productivity, there are negative implications.

Women's domestic burden increases during times of flood. The loss of utensils and other household items is a great hardship for women. They also have to bring drinking water from distant areas as most of the tube well water becomes contaminated. Construction of houses is another post-flood job in which women are involved. Floods can also undermine women's well being because of their dependence on economic activities linked to the home. Loss of harvest and livestock has a high impact on women, as nearly 58% of women rely on cattle and chicken for their cash income.

CONCLUSION

How do poor people live in uncertain and resource-constrained situations? As we have shown in this paper, *char*land inhabitants live difficult lives in eastern India, yet have adapted to this marginal environment and have developed livelihood strategies to ensure their survival. The value of this grounded research lies in locating the place specificness in poor people's livelihood strategies. Cultural rooted ness of people in specific places and occupations continue to play important roles, with the Bihari inhabitants less able to cope with the *char* dynamics than the Bangladeshis. As we saw in this paper, the *char*s have a long and complex history of settlement. People come to live in the *char*s with their limitations, preferences and skills, which they apply in selecting a livelihood in spite of the constraints. The level of diversification of livelihood strategies is high, and while none of the occupations can individually provide a solid subsistence, the overall mix proves to be reasonably adequate for survival, even if on a temporary basis. Social capital plays a significant role in mobilizing kinship networks, not only in migration, but also in building up and consolidating livelihoods once settled. We note the different perceptions and strategies adopted by women and men in creating livelihoods, and highlight that although most women migrated as part of the family the burden of sustaining the family's livelihood falls upon them.

NOTES

1. Biharis are well-known for choosing to migrate to other parts of the country to escape severe poverty in their original homes in north Bihar exacerbated by natural factors such as floods, water logging and shifting river courses. Raja in his 2003 thesis gives an excellent analysis of the Bihari migration in India.
2. A variety of jute which is red in colour and grows without much water and nourishment.
3. Local moneylender.

REFERENCES

Ashley, C. and D. Carney (1999), *Sustainable Rural Livelihoods: Lessons from Early Experience*. London: DFID.

Ashley, S., K. Kar, A. Hossain and S. Nandi (2000), Bangladesh: The Chars Livelihood Assistance Scoping Study, London: DFID, (Draft Paper).

Baqee, A. (1998), *Peopling in the Land of Allah Jaane: Power, Peopling and Environment, The Case of Charlands of Bangladesh*, Dhaka: The University Press Limited.

Bebbington, A. (1999), Capitals and Capabilities: A Framework for Analyzing Peasant Viability, Rural Livelihoods and Poverty, *World Development*, 27(12): 2021-2044.

Bhattacharyya, K. (1998), Applied Geomorphological Study in a Controlled River: The Case of the Damodar Between Panchet Reservoir and Falta. Burdwan, The University of Burdwan (Ph.D. Thesis).

Blaikie, P., T. Cannon, I. Davis and B. Wisner (1994), *At Risk: Natural Hazards, People's Vulnerability and Disasters*, London: Routledge.

Brock, K. (1999), Implementing a Sustainable Livelihood Framework for Policy-Directed Research: Reflections from Practice in Mali, UK: Institute of Development Studies, University of Sussex, (Working Paper 90).

Brock, K. and N. Coulibaly (1999), Sustainable Rural Livelihoods in Mali, UK: Institute of Development Studies, Sussex University, (Working Paper 35).

Bryceson, D. (2000), Rural Africa at the Crossroads: Livelihood Practices and Policies, *Natural Resources Perspectives No. 52*, London: Overseas Development Institute.

Bryceson, D. F., T. C. Mbara and D. Maunder (2003), Livelihoods, Daily Mobility and Poverty in Sub-Saharan Africa, *Transport Reviews*, 23(2): 177-196.

Carney, D. (1998), *Sustainable Rural Livelihoods: What Contribution Can We Make?*, London: DFID.

Carswell, G., A. de Haan, D. Dea, A. Konde, H. Seba, A. Shankland and A. Sinclair (2000), Sustainable Livelihoods in Southern Ethiopia, UK: Institute of Development Studies, University of Sussex, (IDS Research Report).

Chambers, R. (1989), Vulnerability, Coping and Policy, *IDS Bulletin*, 20: 1-8.

Chambers, R. and G. Conway (1992) Sustainable Rural Livelihoods: Practical Concepts for the 21st Century, UK: Institute of Development Studies, University of Sussex, (Discussion Paper 296).

Chowdhury, M. (2001), Women's Technological Innovations and Adaptations for Disaster Mitigation: A Case Study of *Charlands* in Bangladesh, Expert Group Meeting on *Environmental Management and the Mitigation of Natural Disasters: A Gender Perspective*, Ankara, Turkey: 6-9 November.

Cleaver, F. (1998), Choice, Complexity, and Change: Gendered Livelihoods and the Management of Water, *Agriculture and Human Values*, 15: 293-299.

Datta, D. and I. Hossain (2003), *Reaching the Extreme Poor: Learning from Concern's Community Development Programmes in Bangladesh*, Dhaka: Concern (Research Paper).

Deb, U.K., G.D. Nageswara Rao, Y. M. Rao and R. Slater (2002), Diversification and Livelihood Options: A Study of Two Villages in Andhra Pradesh, India, 1975-2001, London: Overseas Development Institute, (Working Paper 178).

De Haan, A. (1998), Migration from Bihar: Understanding Livelihoods in an Historical Perspective, Paper for the Workshop on Migration and Sustainable Livelihoods, Sussex.

De Haan, A. (1999), Livelihood and Poverty: The Role of Migration, A Critical Review of the Migration Literature, *Journal of Development Studies,* 36(2): 1-47.

De Haan, A. (2000a), Globalization, Localization and Sustainable Livelihood, *Sociologia Ruralis,* 40(3): 393-65.

De Haan, A. (2000b), Migrants, Livelihoods, and Rights: The Relevance of Migration in Development Policies, London: Social Development Department, DFID, (Working Paper 4).

De Haan, A., K. Brock, G. Carswell, N. Coulibaly, H. Seba and K. A. Toufique (2000), Migrants, Livelihoods: Case Studies in Bangladesh, Ethiopia and Mali, UK: Institute of Development Studies, University of Sussex, (Research Report 46).

DFID (2002), Better Livelihoods for Poor People: The Role of Agriculture, London: DFID, (Draft Paper – March).

Elahi, K. M., K. S. Ahmed and M. Mufiazuddin (1991), *Riverbank Erosion, Flood and Population Displacement in Bangladesh, Riverbank Erosion Impact Study,* Dhaka: Jahangirnagar University.

Ellis, F. (1998), Household Strategies and Rural Livelihood Diversification, *Journal of Development Studies,* 35(1): 1-38.

Ellis, F. (1999), Rural Livelihood Diversity in Developing Countries: Evidence and Policy Implications, London: Overseas Development Institute, (Natural Resource Perspectives 40).

Ellis, F. (2000), *Rural Livelihoods and Diversity in Developing Countries,* Oxford: Oxford University Press.

Francis, E. 1998. 'Gender and Rural Livelihoods in Kenya.' *Journal of Development Studies* 35(2): 72-95.

Francis, E. (1999), Rural Livelihoods in Madibigo, Multiple Livelihoods and Social Change, University of Manchester: Institute for Development Policy and Management, (Working Paper 6).

Francis, E. (2002), Gender, Migration and Multiple Livelihoods: Cases from Eastern and Southern Africa, *Journal of Development Studies,* 38(5).

Frankenberger, T. R. (1996), Measuring Household Livelihood Security: An Approach For Reducing Absolute Poverty, Paper prepared for Applied Anthropology Meeting, March 27-30, Baltimore.

Ginguld, M., A. Perevolotsky and E.D. Ungar (1997), Living on the Margins: Livelihood Strategies of Bedouin Herd-Owners in the Northern Negev, Israel, *Human Ecology* 25(4): 567-591.

Hapke, H. M. and D. Ayyankeril (2004), Gender, the Work-Life Course, and Livelihood Strategies in a South Indian Fish Market, *Gender, Place and Culture,* 11(2): 229-256.

Hossain, M. I., I. A. Khan and J. Selley (2003), Surviving on their Feet: Charting the Mobile Livelihoods of the Poor in Rural Bangladesh, Paper presented at the Staying Poor: Chronic Poverty and Development Policy Conference, University of Manchester, April 7-9.

IDS (1997), PRA and Gender, UK: Institute of Development Studies, University of Sussex.

ISPAN (1995), Charland Socio-Economic Summery Report, Environmental Study (FAP 16), Dhaka: Flood Plan Co-ordination Organization (Unpublished Report).

Karnath, G. K. and V. Ramaswamy (2004), From Marey to Markert: Changing Faces of Rural Livelihood Systems, In R. Baumgartner and R. Hogger (eds.), *In Search of Sustainable Livelihoods Systems: Managing Resources and Change,* New Delhi: Sage Publications.

Makita, R. (2003), Participation of the Landless Poor in the Rural Economy Through the Non-Farm Sector: A Case Study in Bangladesh, Canberra: Australian National University, (Final draft of research proposal for Ph.D. Thesis).

Marzano, M. (2002), Rural Livelihoods in Sri Lanka: An Introduction of Poverty?, *Journal of International Development,* 14: 817-828.

Masika, R. and S. Joekes (1996), Employment and Sustainable Livelihoods: A Gender Perspective, UK: Institute of Development Studies, University of Sussex, (Bridge Report 37).

McDowell, C. and A. De Haan (1997), Migration and Sustainable Livelihoods: A Critical Review of the Literature, UK: Institute of Development Studies, University of Sussex, (Working Paper 65).

Oberhauser, A. M., J. L. Mandel and H. M. Hapke (2004), Gendered Livelihoods in Diverse Global Contexts: An Introduction, *Gender, Place and Culture,* 11(2): 205-208.

Pearson, H. W. (eds.) (1977), *The Livelihood of Man: Karl Polanyi,* New York: Academic Press.

Scoones, I. (1998), Sustainable Rural Livelihoods: A Framework for Analysis, UK: Institute for Development Studies, University of Sussex, (Working Paper 72).

Siddique, T. (2003), Migration as a Livelihood Strategy of the Poor: The Bangladesh Case, Paper presented at the Conference on Migration, Development and Pro-poor Policy Choices in Asia, Dhaka, 22-24 June.

Swift, J. (1989), Why are Rural Poor Vulnerable to Famine?, *IDS Bulletin,* 20(2): 8-15.

Toufique, K. A. (2001), *Sustainable Rural Livelihoods in Bangladesh,* UK: Institute of Development Studies, University of Sussex.

Toulmin, C., K. Brock, G. Carswell, K. A. Toufique and M. Greeley (2000), *Diversification of Livelihoods: Evidence from Mali, Ethiopia and Bangladesh,* UK: Institute for Development Studies, University of Sussex: Mimeo.

Valdivia, C. and J. Gilles (2001), Gender and Resource Management: Households and Groups, Strategies and Transitions, *Agriculture and Human Values,* 18(1): 5-9.

World Development Report (2000), Washington, D.C.: World Bank.

Yaro, J. A. (2002), The Poor Peasant: One Label, Different Lives. The Dynamics of Rural Livelihood Strategies in the Gia-Kajelo Community, Northern Ghana, *Norwegian Journal of Geography,* 56: 10-20.

26

TRADITIONAL ENVIRONMENTAL KNOWLEDGE OF *BAWALIIS*: A CASE STUDY OF THE SUNDARBANS MANGROVE FOREST[#]

Md. Mustafa Kamal Akand

INTRODUCTION

The Sundarbans Mangrove Forest is one of worlds most unique region, which constitutes the southern part of Gangatic Delta and it extends along the sea facing the Bay of Bengal from the estuary of Hoogly on the west of 24 Parganas (India) to the great river Meghna on the east of Bakherganj District (Bangladesh). The joint flow of the Ganges and the Brahmaputra find their access to the sea. The Sundarbans Mangrove Forest comprises a vast tract of forest and swamp land which occupying an area of 6017 square kilometers or 600,000 hectares represents 60 per cent of the total forest are of Bangladesh and such forms a rich and diverse ecosystem with potential for sustainable natural resources. The forest perhaps got its name from the *sunderi* trees, which grow abundance in this region. The forest lies between 21° 01′ and 23° 01′ north latitudes and 88° 55′ and 89° 58′ east longitudes. It lies about 16 ft above the sea level (GPRB, 1974). The general climatic condition of the SMF is similar to that of adjoining areas of the Bay of Bengal and the forest considerably influence the climate and rainfall throughout the Khulna Division. It enjoys the tropical monsoon weather with warm and humid summer and cool dry winter. It has a fascinating scenic beauty, which may attract many visitors and tourists from home and abroad.

FOREST RESOURCES AND DEMOGRAPHY

The Sundarbans Mangrove forest of Bangladesh possesses numerous species of plants and animals. Different categories of trees, timbers and

This paper was presented at the 3rd Biennial Conference on 'Biodiversity and Quality of Life' organized by Indian Society for Ecological Economics (INSEE) at the Indian Institute of Management, Calcutta, December 18-20, 2003.

source of fuel wood enriched it as one of most resourceful wood storage of this area. A significant number of leaves, bark, shrubs, herbs and many undiscovered nonwoody plants and grasses have made this forest as a unique ecosystem. The dominant species are *sunderi* from which the Sundarbans takes its name, the other species are golpatta, gewa, hantal palm, dhundal and passur occur sporadically throughout the area. *Sunderi*, gewa and keora cover most of the Sundarbans but a great variety of leaves and grasses like, malia ulu, nal are to be found abundantly. The Sundarbans ecosystem is characterized by a very dynamic environment due to tide, flooding, salinity and cyclones. As a result, a remarkable species of animal live here such the flagship species, the Royal Bengal Tiger, spotted deer, monkey, snakes, estuarine crocodiles, dolphins, marine turtles, over two hundred species of fish, crustaceans and moluses. The biodiversity of the Sundarbans mangrove forest also contains a great variety of colorful birds, which attracts the tourists to this region. The colorful bird life along the waterways includes species such as kingfishers (9 species), raptors (38 species), herons, egrets, storks, sandpipers, whimbrel, curlew, gulls, terns, woodpeckers, barbets, shrikes, drongosa, mynahs, minivets, babblers and many others (see Amin and Khan, 2001).

It has always been difficult to trace the real demographic data on the Sundarbans as no population census had been taken even during the British regime. To know the ethnic characteristics and livelihood strategies of the people of this forest one can find a glimpse of historical evidence on the background and occupational statuses in the Bengal Census of 1872, which epitomize the population of the forest as included in the fiscal jurisdiction of Jessore, 24 Parganas and Bakherganj District. W. W Hunter reported that nearly all the inhabitants of the Sundarbans forests were the Bengali Hindus and Muslims. Historically, the forest is inhabited by several groups of people comprising different occupational statuses both from Hindu and Muslims along with a few numbers of other ethnic groups like Munda, Mahato, Maghs, Kols, the Santals and the Dhangors (see Hunter, 1875). Now in the present day, approximately 2.5 million people live in small villages surrounding the Sundarbans. The area provides a livelihood at certain seasons of the year for an estimated 300,000 people, working variously as woodcutters, fishermen, and gatherers of honey, golpatta leaves, hantal and grass (see Amin and Khan, 2001). The local people are themselves dependent on the forest and waterways for such necessities as firewood, timber for boats, poles for house-posts and rafters, golpatta leaf for roofing, grass for matting, reeds for fencing and fish for their own consumption as well as merchandizing throughout the region. Thousands of people, having first obtained their permits from the Forest Department, enter the forest in search of golpatta leaves, hantal, and grass and bee nests. In the coastal area, a large number of people from surrounding villages are employed themselves in garnering of fry throughout the monsoon season and sale these to the commercial scrimp farms. People employed by contractors in the commercial exploitation of *sunderi* and other tree species. Fishermen come in their boats from as far away as Chittagong and establish temporary encampments at various sites along the coast, where they remain until the approach of the monsoon season in April before returning to their homes.

CONCEPTUAL FRAMEWORK

A substantial body of empirical research, scientific literature and documentation on traditional environmental knowledge reflects the growing awareness that there is a legitimate field of environmental expertise known as traditional ecological knowledge. For about a half-century, anthropologists and some animal and plant taxonomists have recognized the accuracy with which various non-western peoples have identified different species. The magnitude of the taxonomic system suggests that the extent of traditional knowledge may be quite profound, and that, indeed, taxonomy is important (as in the biological sciences) as the basis for building extensive systems of knowing about nature.

In the recent social science literature, this aspect of traditional knowledge has to a large extent been documented, more especially, for a large number of so-called foraging (i.e., hunter-fisher-gatherer) peoples, from the tropics to the Arctic (Williams and Hunn, 1982; Ruddle and Johannes, 1991). It is important to note that such traditional ecological knowledge has been found to have management relevance, especially in regard to sustainable use of renewable resources (McCay and Acheson, 1987; Berkes, 1989). It is also important in such endeavors as Environmental Impact Assessment (Freeman, 1975; Craig, 1989; Nakashima, 1990).

It is evident that the need for traditional knowledge research has emerged within cognitive anthropology more recently by reestablishing consciously or unconsciously, the particularistic paradigm proposed by renowned American Anthropologist Franz Boas, who was the first exponent of cultural relativism. He explains human culture as the bundle of traditions in a certain geographic region. To him, all cultures are unique and particular in nature and there is no need to dichotomize culture either superior or inferior. The contemporary participatory approaches to development planning are increasingly stimulating the significance of traditional knowledge. In the domain of participatory development planning, Anthropologists attempt to examine the depth and breadth of indigenous/ traditional knowledge based on ethno-science methodology for discovering the cognitive aspects of people. It rested upon an emic point of view and requires understanding of the symbols and meaning attached to natural phenomena. Chambers (1980:7) argues that,

> We seek to reverse the patterns of development. The thrust of our argument is that 'development from below' is a more productive approach than that

from 'above', and an essential ingredient is indigenous knowledge. Cognitive Anthropology stresses upon eliciting indigenous knowledge as a base for development planning.

Despite the passionate zeal and immense interest about indigenous knowledge, the concept still defies a universal definition. In the rush of western scientific and technological knowledge, the awareness of indigenous knowledge and its depth and relevance is relatively new in world perspective. Until recently, it was considered to be inefficient, and associated practices were thought environmentally damaging and unquestionably out-moded. Long-term cognizance and faith in science and technology (certainly western) promoted the view that indigenous/ traditional knowledge is inferior and in need of change. At the same time, most of the scholars' link indigenous knowledge as politically incorrect label which, rather than esteems to belittle it as not science (Sillitoe, 2000: 7). Different scholars and researchers define the concept from their own point of investigation. Many scholars have made efforts to explain the meanings for Indigenous People and their traditional Knowledge and its importance and complexity. Let us examine some of these key definitions of indigenous/traditional knowledge.

To define the indigenous knowledge, Sillitoe says, 'we are unsure what it is, it is local people's knowledge, what ordinarily folk know (Sillitoe, 1998)'.

In delineating the conceptual development of indigenous knowledge, Chowdhury argues that, 'Each culture has its own internal dynamics that can be developed from within in accordance with its own indigenous tradition and values...' (Chowdhury, 2000: 29).

Although, we are facing ambiguity regarding defining and conceptualizing the term indigenous knowledge, but the following definitions would be supportive to perceive it from different academic point view.

Traditional/Indigenous knowledge is a way of life, an experience-based relationship with family, spirits, animals, plants, and the land, an understanding and wisdom gained through generations of observation and teaching that uses indirect signals from nature or culture to predict future events or impacts.

This knowledge derives from, or is rooted in the traditional way of life of aboriginal people. Traditional knowledge is the accumulated knowledge and understanding of the human place in relation to the universe. This encompasses spiritual relationships, with the natural environment and the use of natural resources, relationships between people, and is reflected in language, social organization, values, institutions and laws (Legat, 1991).

Indigenous knowledge (IK) is the local knowledge – knowledge that is unique to a given culture or society. IK contrasts with the international knowledge system generated by universities, research institutions and private firms. It is the basis for local-level decision making in agriculture, health care, food preparation, education, natural-resource management, and a host of other activities in rural communities. There is consensus amongst scientists using various terms that such knowledge: i) is linked to a specific place, culture or society; ii) is dynamic in nature; iii) belongs to groups of people who live in close contact with natural systems; and iv) contrasts with "modern" or "Western formal scientific" knowledge" (Warren, 1991).

The ecological dimension to traditional systems of knowledge is usually taken to refer to a people's medicinal, technical, and ritual uses of plants, animals, and rocks; to place names and occupancy of territory; or to the spiritual, cosmological, and relational aspects to the various presences (animate, inanimate, present or past) in an environment. IK is local environmental knowledge (knowledge of plants, animals, soils and other natural components) with practical applications, rather than the more encompassing sense of IK associated with environmental philosophies or world-views, or even ITK (indigenous technical knowledge) in its wider sense. However, we accept that such practical, technical and empirical knowledge is characteristically embedded, linked to, and informed by these broader understandings. It is assumed that most readers will already be persuaded that indigenous environmental knowledge (hereafter IK) can hardly be ignored in development contexts, as well as having applications in industry and commerce.

Traditional Ecological/Environmental Knowledge (TEK) is a body of knowledge built up by a group of people through generations of living in close contact with nature. It includes a system of classification, a set of empirical observations about the local environment, a system of self-management that covers resource use. With its roots firmly in the past, traditional knowledge is both cumulative and dynamic, building upon the experience of earlier generations and adapting to the new technological and socio-economic changes of the present (Johnson, 1992).

To Studley "... a body of knowledge built by a group of people through generations living in close contact with nature. It includes a system of classification, a set of empirical observations about the local environment, and a system of self-management that governs resource use"(Studley, 1998).

According to Grenier "... traditional ecological knowledge can be defined as a cumulative body of knowledge and beliefs, handed down through generations by cultural transmission, about the relationship of living beings (including humans) with one another and with their environment. Further, traditional ecological knowledge is an attribute of societies with historical continuity in resource use practices; by and large, these are non-industrial or less technologically advanced societies, many of them indigenous or tribal"(Grenier, 1998). It is a structure of values, stories, language, and social relations. It is an experience-based relationship with family, animals, places, spirits, and the land. It is a worldview. Because it is experiential, each group will have a tradition that is, to a greater or lesser extent, different from other groups. While it may be convenient to speak of indigenous traditional knowledge, it is important to understand that such knowledge is not a single homogenous body. In fact, many indigenous peoples would say that their traditional knowledge is only accessible to its fullest extent by those who understand the language of their people.

From the multidimensional perspectives, we can define traditional environmental knowledge as a body of knowledge and beliefs transmitted through oral tradition and first-hand observation. It includes a system of classification, a set of empirical observations about the local environment, and a system of self-management that governs resource uses. Ecological aspects are closely tied to social and spiritual aspects of the knowledge system. The quantity and quality of TEK varies among community members, depending on gender, age, social status, intellectual capability, and profession (hunter, spiritual leader, healer, etc.). With its roots firmly in the past, TEK is both cumulative and dynamic, building upon the experience of earlier generations and adapting to the new technological and socioeconomic changes of the present.

In this paper traditional environmental knowledge has been perceived as the cumulative body of knowledge, beliefs and practices of *Bawalis*, about the environment on which they live. This knowledge has been practiced by their earlier ancestors over the history and gradually it is handed down to the present generation as a cultural pattern of this group. To maintain the livelihood, *Bawalis* always practicing a kind of interaction with the surrounding environment on the basis of local knowledge, which is unique and deeply rooted into the historical context of the Sundarbans Mangrove forest. Without any major changes, *Bawalis* traditional environmental knowledge is being going after very firmly, and this knowledge shapes up their social organizations, rituals, values, institutions and laws in using the natural resources continuously.

STUDY AREA AND DATA SOURCES

This paper is an out come of the field-based data collected from three villages of Atlia Union of Shamnagar Upazila under Satkhira district, Bangladesh. The approximate area of Shamnagar upazila is about 759.94 square miles and its population is more or lees 2,65.004 according to 1991 Census. The Shamnagar upazila situated within the area of Sundarbans forest and major part of its land representing the forest area. The selected villages are Choto Kupat, Boro Kupot and North Atlia. It is an important landing spot of forest resources. Most of the people of these villages are directly involved with the Sundarbans resource collection and utilization. It is also well known to the researcher that a significant number of samples are available which the present study deals. 50 households (320 population of different age group and sex) have been identified randomly for data collection considering their involvement with the exploitation of forest resources.

Table 1
***Bawalis* Socioeconomic Background in Three Study Villages**

Socioeconomic Variables	*Frequency N = 50*	*Percentage %*
Age Group		
20-44	33	66.00
45-49	09	18.00
50-59	04	08.00
60 ≥	04	08.00
Education		
No schooling	29	58.00
Primary (i-v)	17	34.00
Secondary (vi-x)	03	06.00
Passed SSC	01	02.00
Family Types		
Nuclear	31	62.00
Extended	19	38.00
Family Size		
Small (1-4)	13	26.00
Middle (5-8)	28	54.00
Big (9 ≥)	09	18.00
Average Size (6.4)		
Land Ownership		
Landless (Less than 0.32 acres)	42	84
Marginal (less than 5 acres)	08	16.00
Total	**50**	**100**

Figure 1
Sundarbans Mangrove Ecosystem

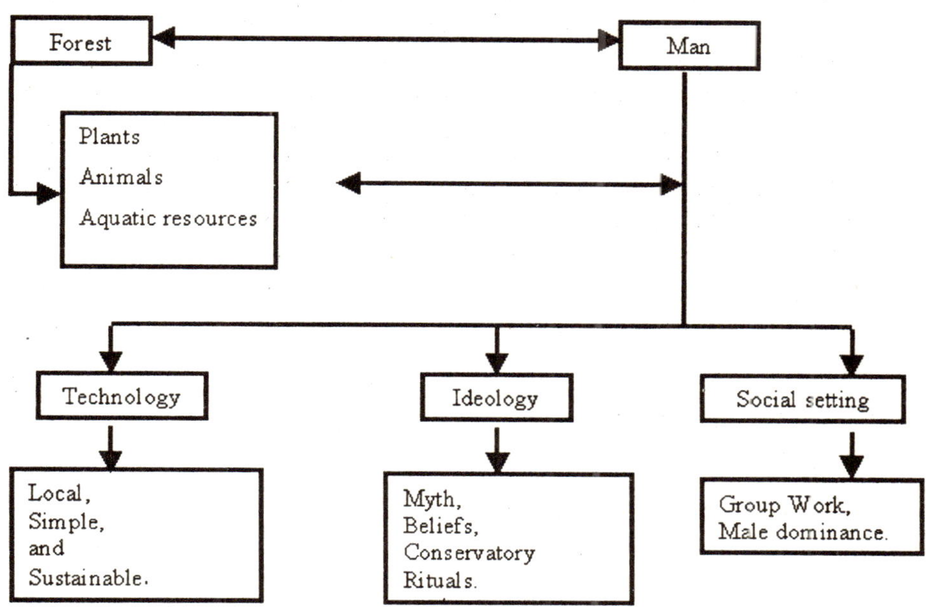

BAWALIS ENVIRONMENTAL KNOWLEDGE

The natural resources provide the basis of human actions; where as, human potentials for being organized, knowledge, technology and cognition provide another resource base. As we know, several alternatives levels of production are possible within a natural condition, but the selection procedures depend on how the individuals evaluate and take decisions in terms of the existing knowledge and experience along with the structure of local opportunity cost. According to Barth, Nature as a potential source, offers a certain amount of energy that can be exploited. The exploitative activities are culturally conditioned (Barth, 1981: 181).

In order to identify the *Bawali* adaptation strategy, I would like to illustrate different sphere of environmental interaction, which is clearly designed traditionally and circumscribed by the natural environment of Sundarbans Mangrove forest (see Figure 1). To document the major interaction with the nature, I have traced three major spheres of activities governing by traditional knowledge, which are very much associated with their livelihood and survivals. These are as follows.

Harvesting Leaves

Different species of non-wood forest products from the Sundarbans forest play a major role in the lives of hundreds of thousands of people who live near its boundaries. These products are harvested for both subsistence and commercial purposes and represent an important source of livelihood for the surrounding villagers. Among the villagers under study, almost every household is engaged in collection of different leaves and non-wood products. Round the year they collect wood and non-wood materials from the forest, but winter is more viable to harvest, as the wild animals remain less active that time. Golpatta is a regal palm that grows in monotypic stands along the banks of the rivers and canals that bisect the dense mangrove forest of the Sundarbans. From December to May, the *Bawalis* collect tones of fronds from throughout the forest. The fronds are sold at several trade depots in the adjoining districts and used as a traditional roofing material throughout the southern and western part of the country. The fruits are used to make local wine. Other species are also important as well. Hental is a palm used in construction as rafters, fence and framework. It is not as popular as golpatta and only harvested in smaller quantities. On an average, 3000 tones per year. A great variety of grasses are annually harvested from Sundarbans, which includes malia, nal and ulu. Malia is used for making mats, nal for baskets and fence and ulu for thatch. The *Bawalis* traditionally harvesting and preserve golpatta and other non-wood leaves by using simple, traditional tools, vehicles and methods. For instance, to collect golpatta from the embankment of the rivers and canals, they use sharp cutting edge taking in mind always the sustainability of the species and cut it in an angular shape on a certain height, from which buds can grow naturally for further harvesting. Before leaving the harvesting spot, they do a conservatory ritual and planting a new bud in the ground, as they believe in a myth that, *banbibi* may destroy them if they don't perform this traditional ritual. After a series of piling up in a small boat, they later deposit the leaves to a huge *panshi* boat. *Bawalis* follow the natural law of blow tide and ebb tide to make their boat journey easier. They follow a particular rule of piling known as *jagh*, which keeps the leaves softer and fresh. These leaves never dried in the sunlight, because the leaves may break if these are dried in the heat of the sun and may lessen its quality.

Collection of Honey

As the great deltaic Sundarbans is abundant with thousands of trees, that help to produce a lot of nectar and pollen. For many thousands of years, this rich floral resource has been used by a species of bee called Giant Honey Bee. Many thousands of colonies of this magnificent honeybee congregate within the forest to take the advantage of the millions of millions flowers that provide nectar and pollen for their colony development. An exceptional colony of giant honeybees can produce up to 30-40 kgs of honey, although an individual colony more frequently has 5 to 10 kgs. The honey is stored in the upper part of the wax comb, which be large as

1.5 meters across and I meter in depth. Generally, honeybees build their large nests high about 30-40 meters on large trees limbs in other Asian forests, but in Sundarbans, because of the nature of mangrove forest which does not have many tall trees, the bees must build their nests at relatively close to the ground, only 1 to 5 meter high. Generally, March, April and May are the most productive period of honey collection. Before the beginning of monsoon, people known as *Mawalis* (engaging in honey collection though the person is same engage in collection of leaves and wood) enter into the forest to search for giant honeybee colonies. They always follow the lunar calendar to get the best result of harvesting honey. Expedition for honey collection is designed on the basis of lunar calendar. From their experience, they know that honeybees preserve nectar in their colony for a certain period of time and bees consume the nectar themselves in the new moon. Therefore, harvesting must be completed 4/5 days before the new moon. Entering into the forest, they can correctly identify the productive colonies. It may be pointed out that, mawalis do not work as individuals, but rather form small group of between 5-10 men who work cooperatively in their hunting activities. The honey collector uses small boats, as the tidal mangrove ecosystem possesses no roads connected with surrounding locality. They roam many kilometres away from their home village and must spend many days (2 weeks at least) taking sufficient food, water and other necessities with them to sustain themselves during the arduous spent in the forest seeking the honeybees. After a long searching, when they find a hive of honeybees, they explore the news by producing a kind of sound colloquially *wa, wa* and seek the advice from expert member of the team with great interest. The leader confirms it for hunting. First of all, *Mawalis* try to dislocate the bees from the colony by *kadu* (a stick with a bundle of dry and wet leaves on the head of the stick which produce smoke). Then they carefully separate the part of colony containing honey and dispose it on the basket. This action is always done at a higher sincerity as to avoid the destruction of bees and colony forever. They believe that a colony can be used for further hunting if the queen bee and her surrounding associate members remain with her in the colony and protect from despoilment. Generally, 7-10 colonies are harvested in day. After the hunting is over, they return back with their collection to their boat and then crushed to extract the honey and finally preserve into the plastic or pottery jar.

Fishing

Collection of wood, honey along with leaves and grasses are not only the major activities to maintain their livelihood of Sundarbans forest people, as we know the deltaic Sundarbans also a significant source of aquatic resources abundant with hundreds of fish species. People from surrounding villages are largely engaged in fishing. Over the year, they collect fry; catch huge numbers of different fishes in different sizes from the coastal area. In the monsoon, they catch fish in large scale and sale them directly to the nearby trading depots. But in the winter, the villagers generally collect the fish to make them dry and sale it over the year. To complete the total process of drying and preserving the fishes, they practice traditional knowledge based actions. At first stage, fishes are classified on the basis of species and size and then they spread hot water on them, which protect the fishes from infectious germs and help them to be dried easily. No formal chemical is used for preserving the dry fishes.

ASSESSING VULNERABILITY AND MITIGATION

People live in Sundarbans forest environment always facing severe difficulties and risk unlike the plain Landers. They frequently encounter different species of ferocious animal both aquatic and non-aquatic, poisonous plants, uncertain weather, fear of robbery and so on. To ensure the continuity of their livelihood, *Bawalis* have traditionally developed a sustainable system of risk assessment and mitigation. Nature directly teaches *Bawalis* in assessing the environmental vulnerability as well as the necessary mitigation process. Knowledge on weather and danger (created by tiger, snakes, wild pigs, poisonous plants, crocodiles and sharks) is an essential part of the *Bawalis* livelihood. Both in the forest and sea, they apply their traditional knowledge to ease the total adaptation process. Proper assessment makes them safe and winner. They can predict the unfavourable condition of the forest and weather by judging the behavior of different animals like birds, bees, monkeys and deer. For example, when they observe that the nests of bees are in unnecessarily high, birds nests are dramatically low, both indicates the bad forecast. In the forest, they can also identify the presence of tiger and other ferocious animals by examining their footprint on the ground. The abnormal behavior of monkeys and deer also aware them and say something only they can read about the nearest danger of tigers. They also use different local medicinal plants and herbs in different cases. When snakes bite a person, they test it by feeding neem leaves to the victim. They also use mustered oil to reduce the poison. If they affected by the roots of *sunderi*, which causes serious damage on the skin, *Bawalis* use crushed snail, mustered oil, turmeric and shrubs extract. They also wear gumboot to prevent themselves from this kind of danger.

The fishermen generally ignore the formal weather forecast, because they traditionally can predict the condition of weather forecast on the basis of the following criteria. These are:

(i) The abundance of fishes increases near the coast.

(ii) Increasing salinity.

(iii) Warmness of water.

These above indicators of cyclonic weather play a two-way role among the people; they remain closer to the coast, enjoys the catching fishes in abundance in one hand. And, this cycle of environmental interaction helps them to survive on the other. To predict the depression or cyclone, there are some other indigenous indexes of weather also helping them to cope with the subsequent threat and danger created by the nature. For instance, the direction of wind blow from southeast corner is a clue of storm. Red cloud and an appearance of rainbow signal the deep depression in the sea.

CONCLUSION

People around the world, have long recognized the ecological values of mangroves. Though, mangrove forests are frequently cleared to accommodate expansion of human populations, to support agriculture and aquaculture, to provide room to install cultivated shrimp ponds and to convert them for other human uses, This practice has contributed to the decline of thousands of acres of mangroves around the globe. This widespread habitat conversion contributes to forest fragmentation, concomitant loss of species both animals and plants, and destabilization of mangrove-dominated shorelines. If the ecology of natural resources becomes better understood, it can equip the user, stake holders, policy planners and conservators more effectively to develop strátegies for successful restoration and sensible, sustainable harvesting of appropriate mature mangrove resources. The present study explores some unique instances of traditional knowledge regarding sustainable forest management, which can help us to take in account, the significance of local initiatives and peoples participations. This paper on *Bawalis* also reveals the reality that; people of Sundarbans developed a local system of interaction with their nature. In order to adapt to the environment they have been practicing a different mode of livelihood. It is obvious that, local knowledge has developed by the people themselves to mediate man and environment interaction. It is locally appropriate, respect for nature, flexible, restraint in resource exploitation; production is for subsistence needs only and what is needed for immediate survival is taken only from the environment for a long period of history. This environmental knowledge has successfully been maintaining a state of equilibrium with the nature and ensuring the continuity of *Bawalis* livelihood.

REFERENCES

Amin, M. N. and Khan (2001), A Primer on the Nature and Management of Resources of the Sundarbans, *Grassroots Voice: A Journal of Indigenous Knowledge and Development*, Vol. IV: 87-92.

Barth, F. (1981), *Process and Form in Social Life: Selected Essays of Fredrich Barth*, Vol. I, Routledge Kegan Paul Ltd, London.

Berkes, F. (ed) (1989), *Common Property Resources: Ecology and Community Based Sustainable Development*, London: Belhaven Press.

Chambers, R. (1980), Introduction. In D. Brokensha, D. M. Warren and O. Werner (eds) *Indigenous Knowledge Systems and Development*. University Press of America, Washington.

Craig, D. (1989), *Resolution of Conflict in Australian Water Management: Aboriginal Interests and Perspectives*, Centre for Resource and Environmental Studies, Australian National University, Canberra.

Chowdhury, A. F. H. (2000), Indigenous Knowledge System: A Development Paradigm in Anthropology In Niaz. A. Khan (eds.) *Indigenous Knowledge and Practice in Bangladesh*, BARCIK, Dhaka.

Freeman, M. M. R. (1975), Assessing Movement in an Arctic Caribou Population, *Journal of Environmental Management*, Vol. 3: 251-257.

GPRB (1974), *Population Census of Bangladesh*. District Census Report-Khulna, Dhaka: Bangladesh Bureau of Statistics.

Grenier, L. (1998), Working With Indigenous Knowledge: A Guide For Researchers. IDRC: Ottawa, Canada.

Hunter, W. W. (1875), *A Statistical Account of Bengal*. Vols. I & II, London.

Johnson, M. (ed.) (1992), LORE: Capturing Traditional Environmental Knowledge. Dene Cultural Institute and IDRC, Ottawa.

Legat, A. (ed.) (1991), Report of the Traditional Knowledge Working Group, Government of the Northwest Territories, Yellowknife.

McCay, B. J. and J. Acheson (1987), *The Question of the Commons*. Tucson: University of Arizona Press.

Nakashima, D. J. (1990), *Application of Native Knowledge in EIA: Inuit, Eiders and Hudson Bay Oil*, Report Prepared for the Canadian Environmental Assessment Research Council, Hull.

Ruddle, K. and R. Johannes (ed.) (1990), *The Traditional Knowledge and Management of Coastal Systems in Asia and the Pacific*, Jakarta: UNESCO (2nd edition).

Silitoe, P. (1998), The Development of Indigenous Knowledge: A New Anthropology, *Current Anthropology*, Vol. 39, No. 2: 223-252.

Silitoe, P. (2000), Some Comments on Science, Indigenous Knowledge and the Poorest of the Poor in Bangladesh, in Niaz. Khan (eds.) *Indigenous Knowledge and Practice in Bangladesh*, BARCIK, Dhaka.

Studley, J. (1998), *Dominant Knowledge Systems and Local Knowledge*. Mtn-Forum On-line Library Document, Accessed August 2001.

Warren, D. M. (1991), *Using Indigenous Knowledge in Agricultural Development, World Bank Discussion Papers*, No. 127, Washington, D.C.: The World Bank.

Williams, N. M. and E. S. Hunn (eds) (1982), *Resource Managers: North American and Australian Hunter-Gatherers*. Boulder, CO: Westview Press.

27

HISTORICAL RELEVANCE OF JOINT FOREST MANAGEMENT PROGRAMME AND THE KEY ELEMENT OF ITS SUSTAINABILITY: EVIDENCE FROM WESTERN MIDNAPORE DIVISION OF WEST BENGAL

Debnarayan Sarker and Nimai Das

INTRODUCTION

In the context of Indian forestry, several strands have gone to make up the present emphasis on community involvement in forest protection. Joint Forest Management (JFM) can be seen to emerge as the latest in the history of policy changes, as well as one more in a series of attempt to create a new relationship between the state and the community in terms of forest management .The old custodian forest management system developed were rendered ineffective in 1950's and 1960's due to various reasons, the major reason being traditional emphasis on production of commercial wood and disregard for local need. Against this old custodian forest management system the local communities in different parts of India have mobilized repeatedly from long past to protect their resources from manipulation from outside groups. The emergence of new community forest management system in west Midnapore division in West Bengal, like some other parts of rural India, is grounded historically in tribal and peasant resistance movements.

The paper is organized as follows: section I presents the historical perspective of the study, section II provides a short review of the relevant literature, section III covers the data set and findings, and conclusions are contained in section IV.

HISTORICAL PERSPECTIVE

In the pre-colonial period, western Midnapore was covered by dense Jungle tracts (dense forest). While patches of forests, particularly along river plains, had been cleared for agriculture, much of the area was wild and remote. Western Midnapore, primarily populated by Santal, Bhumij, and Mahato tribals, with some low-caste Hindus, included the police stations (Thana) of Garbetta, Binpur, Gopiballavpur, Salboni, Sildha and Jhargram. Prior to the colonial era, while this area was nominally under Mughal control due to the inaccessibility of the area, little attempt was made to extract revenues or exert political authority. Forest-dwelling communities of this area could resist incursions into the area. Their superior knowledge of the jungle and their hunting skills made them an effective guerrilla force. Some Bhumij communities gained the reputation as CHAURS (robbers), from their aggressive raids into the plains. Many local rajas (kings) and zamindars (land-lords) preferred to leave them alone and not attempt to extract taxes from them, rather than to enter conflicts with jungle people (Poffenberger, 1995).

Many tribal communities that maintained forest oriented self-sufficient economics were best able to obstruct outside political domination. They alternatively protected their political autonomy and forest resources. The Santal and Bhumij tribal communities of forest inhabitants practiced shifting (swidden) cultivation as well as hunting and gathering forest products[1]. Much of their diet was provided from wild fruits, roots, herbs and the nutritious flowers and fruit pulp of the Mahua (Madhuca) tree, making them less dependent on agriculture. Tribal villagers were also actively engaged in trade in firewood, silk, resin, deer and buffalo, horns, wax, honey, bark fabrics, lac, medicine and charcoal[2].

Along with the survival needs of the forest communities, cultural values of them also retained health, fertility and prosperity of the forest. During the pre-colonial period, and up to the present, the belief systems of the forest communities of this area were strongly grounded in the worship of nature. Religious festivals are tied to both the agricultural cycle and the flowering and fruiting of the forest trees. The Santal New Year, for example, begins with the blossoming of the sal (shorea robusta) tree in March. This links in tribal beliefs[3] helped the regeneration of natural forest.

In 1760, the district of Midnapore was transferred to the East India Company by Mir Qasim, making it one of the first districts in India to be brought under British rule[4]. During the late eighteenth century the British sent military expeditions into this area in an attempt to extend their authority and extract land revenues. The forest chieftains and tribal communities resisted, ambushing British forces and harassing them whenever possible. Local zamindars also resisted the imposition of colonial authority, refusing to pay their taxes, organizing their paik militias to resist, and falling into arrears on their taxes. In 1798, widespread violent resistance disrupted revenue collection activities in the Midnapore area, forcing the Company to restore lands to hereditary chiefs that had been put up for sale for failure to pay taxes.

Through superior force, however, the British did gradually succeed in extending their control in the area through the nineteenth century. As this process continued, the British empowered a new class of zamindars to control and tax local forest communities, encouraging them to open forest land for cultivation. In order to meet their tax obligation, zamindars were anxious to bring in tribal and peasant cultivators to clear forest and convert it to agricultural land. The tribal communities of this area resisted the imposition of the taxing and conversion system through a series of armed revolts[5]. The first, referred to as the Chuar Rebellions, lasted from 1767 to 18005. Tribal guerrillas were so effective that 'even as late as 1800, after nearly forty years of British occupation, a collector reported that two thirds of Midnapore consisted of jungle, the greater part of which was inaccessible'[6]. Yet gradually the Company succeeded in strengthening its control, despite subsequent revolts by forest people, such as the Naik Revolt (1806-16). Under the Permanent Settlement Act by 1866, 1369 zamindari estates had been established in Midnapore, and given the absolute ownership of agricultural and forest lands as long as they paid government revenues.

The process of forest clearing for agricultural land conversion had sweeping ecological implications, especially for river systems and soil conditions. Removal of forest cover allowed torrential monsoon rains to wash away the shallow topsoils, leaving an exposed laterite hard pan that made farming virtually impossible in many areas. Traditional forest based industries like tusar silk, indigo and endi declined dramatically, as did the population density of this area as the forest was cleared.

The pressure on the forest grew further by the 1860s as the growing railway system demanded immense quantities of sal logs to provide sleepers for rail bed. Commercial demand for timber accelerated forest cutting, and raised the value of forestlands. Timber merchants rushed in, even before the rail lines opened and began leasing or purchasing large tracts from the Midnapore Zamindary Company and other zamindars.

In early 1855, six to seven thousand Santal tribals from, Birbhum, Bankura, Chotonagpur and Hazribagh began meeting for organizing resistance in response to their growing marginalization. On July 16, 1855 some ten thousand tribals, under the messianic leadership of four Santal brothers stood their ground firmly and fought with bows and a kind of battle-axe in a battle near Pirpaiti[7]. The revolt collapsed eventually after half their members were reportedly killed. Despite their defeat the HUL Rebellion (as it is known among the Santal) profoundly influenced the ideological development of many Santal communities[8], and lives on in the songs and oral traditions of the tribal people of this area. Throughout the later part of the nineteenth and first half of the twentieth century, many forest communities on this area became increasingly indebted to moneylenders and tax collectors, causing widespread mortgaging and loss of their agricultural lands. Though the alienation of private lands was an important element in the impoverishment of tribal and low-caste communities, so too was the loss of cash and kind income from forest-based activities as the forests were cleared.

The forest policy of post-colonial India continued on the colonial path of commercialization and reductionism, and with it continued people's resistance to a denial of their basic needs, both through alienation of rights and through ecological degradation (Shiva, 1999). Despite the populist government in West Bengal after independence the old custodian system of commercial forestry, which disregarded local need also prevailed in western Midnapore. Throughout 1969 and 1970, some forest communities in this area were discontented with the Forest Department and its policies of providing elites and contractors with low-cost resource exploitation leases. Forests were logged of timber trees and bamboo and villagers lost the raw materials they required for their subsistence and commercial needs. The Dom tribals in particularly were upset by the high prices and fuel wood scarcity experienced by potters, blacksmiths, and other caste groups also increased antagonism towards the forest department and those who acted as contractors for them.

While successful examples of joint forest management were beginning to emerge in Arabari Hill in Midnapore district during the early 1970s, throughout the decade they remained isolated cases with little effect on routine forest management within the state. Recognizing the success of Arabari and a few other villagers where management agreements with forest communities were being made, some senior forest officers like Dr. Ajit Kr. Benerjee, began to encourage field staff to pursue similar negotiations in wider areas throughout the southwestern part of the state. By formulating agreements that responded to the economic needs of forest communities, new incentives were created among villagers, which resulted in the emergence of effective controls on forest exploitation. The emergence of Chingra Forest Protection Committee, Chandana Forest Protection Committee, Harinakuri Forest Protection Committee are the examples of decentralized nature of forest protection group formation in Midnapore district. This early experience demonstrated that opening communications with forest communities could effectively reduce conflicts between the forest department and forest user groups. Forest officers were able to identify terms for effective management partnerships through discussions. In some communities village men formed volunteer patrols. Village volunteers warned people who were found cutting green wood or grazing animals. Repeat offenders from the participating villages were fined, and outsiders were turned over to forestry field staff.

The experience of some forest protection committees (FPCs) in this area illustrates the way in which village leaders like Mahadev Munda Singh of Chingra FPC, Lokhun Sahu of Chandna FPC and Joti Naik of

Harinakuri FPC were able to work with field staff and other neighbouring communities to identity forest areas for protection and reach agreements, while turning away outsider users. It appears that the ability of local communities to take the lead in defining management territories was a key to the success of the programme. Although field staff helped facilitate this process by encouraging group meetings and authorizing community protection activities, frequently successful FPCs took the initiative in organizing themselves and establishing operational controls over forest access. Most confrontations came into effect during the first and second year of protection after which restriction and right of the protection committees were generally recognized by outsiders.

Despite its early success, these early achievements were limited to small forest tracts of this area. The JFM in this area, however, was largely effective among most of the forest communities neighbouring the forest when West Bengal Government Orders were issued on 12th July, 1989 to formalize the FPC wherein the duties and responsibilities of the FPC members, usufruct and other benefits to which they would be entitled had been laid down. The departmental appeal to tribal communities to protect forest resources and its willingness to empower them apparently coincided with a growing desire among these communities to take environmental action. It was easier for communities to mobilize because the West Bengal programme did not require complex registration and budgetary allocations processes for communities to take action, but rather presented communities with a straightforward opportunity (protect the local forest and enjoy the benefits). As each community began protection activities, it influenced the behavior of neighbouring villages. Without necessary waiting for the forest department to take action, villagers were forced to negotiate and discuss management issues and needs with one another (Poffenberger, 1995). It is this community based 'chain reaction' or catalytic effect that is apparently a driving force behind the rapid emergence of localized access controls on state forest lands in southwest Bengal[9]. As per the latest State Forest Report published from Directorate of Forest, Govt. of West Bengal on December 2001, the total number of FPC in west Midnapore division was 542, the highest in number among all divisions in West Bengal. Thus the historical events outlined earlier in this discourse suggest that the communities in this area have mobilized repeatedly over long past to protect their resource rights from manipulation by outside groups. This study indicates that the emergence of new wider community forest management systems with a straightforward opportunity is grounded historically in tribal and forest communities' resistance movements. In many parts of rural India, pockets of dis-empowered people have organized repeatedly to struggle for their survival as local elites, moneylenders, tax collectors, and the state increasingly capture their resource base. In the past, each time the movement was crushed or collapsed, after some time it would reemerge. The people of western Midnapore in West Bengal represent a classical case.

REVIEW OF STUDIES

While empirical evidence across the world now confirms that community-based regimes are a viable option for the management of local common property resources (Baland and Platteau, 1996; Berkes, 1989; Bromley, 1992; Correa, 1999;Lama and Marlene, 2002; Martin, 1992; McCay and Acheson, 1987; Naik, 1995; Saxena and Sarni, 1999; Singh, 1994 and 2001), the theoretical literature has developed an understanding of the mechanism which make these regime work. The early research was pessimistic about the possibilities of group management, arguing that individuals would not face the full costs of their resource use under such regimes, and would have an incentive to over-exploit the resources. This was the 'free-rider' problem and the only way to solve this problem, according to their suggestion, was to create private property rights in the resource, or to regulate resource use by coercion through the state. This early theoretical tradition was inconsistent with the empirical evidence on group management regimes, and it was subsequently pointed out that the free-rider problem was a characteristic of the 'unregulated commons' or 'open access'. This was distinct from 'common property', a regime in which a clearly defined group co-operated to manage a resource (Ciriacy-Wantrup and Bishop, 1975). It is no longer theoretically disputed that individuals can coordinate their actions and participate in collective regimes.

As better management of common property resources, such as forests, is considered vital for poverty alleviation, sustainable development with equity, ecological stability and preserving bio-diversity, there is a constant search for alternative approaches necessitated by the fact that the usual options of state or market mechanism are not advocated due to their inherent inadequacies. There is evidence to suggest that privatization of such common property resources would enhance inequity of denying access to underprivileged classes of the community (Karnath, 1996). But the survival of community needs of poor communities need to be recognized on a priority-based as pillars for strengthening their community participation (Mukherjee, 1995). Some authors produced general lists of conditions, which facilitate successful community based resource management (McKean, 1992; Wade, 1998; Baland and Platteau, 1996; Ostrom, 1990). Eleven factors are commonly identified in this literature as creating conditions which are more conducive to local-level management[10]. But expectations of immediate returns via wages and incomes from sale of old plantation and local consumption need to fill the requirements of fuelwood, fodder, minor forest produce and small timbers seemed the most important factors motivating massive local peoples, participation for protection and development of forests (Mukherjee, 1995; Naik, 1997;

Saxena and Sarni, 1999). The present process and basis of developing VFPCs/FPCs will need considerable strengthening if these have to be developed into robust and confident village institutions. The first important step in this direction is of ensuring that FPC members have clear entitlement to all types of forest produces from redefined JFM areas for meeting their bonafide requirements, with no removal of produce from the area unless it is surplus to local consumption needs. Moreover, the present policy of selecting the most degraded land for planting would have to be modified in favour of greater emphasis on regeneration and less on planting (Saxena and Sarni, 1999; p. 213).

But long-term gains hardly matter to people who are facing major problems of livelihood-food security. In depressed areas with higher level of poverty day-to-day existence comes top most on their agenda. The most urgent community need under JFM at Arjuni in Midnapore district, West Bengal is that of alternative means of livelihood during lean season with agriculture being the mainstay in peak season. The Arjuni experience in JFM shows that unless survival needs of food and livelihood are met, participation of natural resources management would always remain threatened (Mukherjee 1995; p. 3132). This experience goes a long way to show that survival needs are of prime importance and can easily destabilize community rights and benefits to resource management. Any JFM, which does not recognize the significance of creating strategies for sustaining livelihood – food security – at the local level, has a doubtful future (ibid, p. 3132). The findings of Naik (1997), based in two case studies in Gujrat state, can also help identify the factors critical in making the JFM successful and if there are controllable, such as market development and share of the local people in forest produce, necessary steps may be taken to suitably change the factors in order to enhance the chances of success of JFM. The present JFM needs to be remodeled in terms of livelihood benefits which are immediate and less commercial (Mukherjee, 1995; p. 3132). To this end, the forest policy of Government of India, 1998, recognized the need to fulfill the requirements of fuelwood, fodder, minor forest produce and small timber of rural and tribal people, and emphasized the need to create massive people's movement for protection and development of forest.

But concerning to the benefit-sharing arrangement between states and forest communities and the management of forest polices differ between states within the country. Orissa's NTFP policies are among the most regressive in the country, guided by the principle that all forests and forest produces are state property; the NTFPs gatherers access to income from NTFPs are severely restricted due to existing policies. The Vana Samrakshan Samiti (VSS) gatherers get only wages for collecting the products. Thus even in jointly managed forest lands, the people, who are supposedly co-managers, are treated as mere hired labourers whose earnings are based on the minimum wage rate and not treated to the value of the products (Rao, 2001; p. 261). In JFM, and also in the recent NTFP policy in Orissa, no steps have been taken to address the real problem of NTFP's gatherers. The issues affecting the livelihoods of forest dependent poor women and men remain un-addressed, and any talk about community participation in management of forests and community rights remain meaningless (*ibid*, p. 262).

The field context in which JFM is implemented brings together two principal players: the village community and the forest bureaucracy. Forest bureaucracy, in many cases, enforces rather than abdicates its power influencing adversely the relationship between forest communities and forest department. Benefit-sharing arrangement in many states have also been specified in advance, without asking villagers whether they, in fact, want to harvest their timber, and how they would like the proceeds to be distributed. In most states, the committees are simply registered with the forest department. Only in five states (Gujrat, Haryana, Rajasthan, Arunachal Pradesh and Karnataka) do the VPCs have a legally independent existence as co-operatives registered under the Co-operative Society Act at least formally (Jeffery and Sundar, 1999; p. 45).

In most cases the forest department also reserves the right to dissolve committees if they perform unsatisfactorily, or at least deny them the shares expected (Poffenberger and Singh, 1996; p. 71, Jeffery and Sunder, 1999; p. 45). As happened in the Panchmahals in Gujrat, parallel processes exist, refusing to register existing committees, specially if the forests they have been protecting have now turned from degraded into good forests (Agarwal and Saigal, 1996; p. 7); or attempting to latch onto existing committees while simultaneously denying the legitimacy of rules they have been framed earlier and which do not fit into state resolutions (Sundar *et al.*, 1996). Moreover, planners, policy makers and forest bureaucracy, in many cases, fails to develop appropriate strategies to successfully involve the poor (including women) who depend much more on the forest as they have little access to alternative source of biomass in community forestry. It has been well documented that community-based projects and programmes (specially in India and Nepal) ignore women and the poor (Agarwal, 1997; Hildyard *et al.*, 1998; Hobley, 1996; Joekes *et al.*, 1996; Locke, 1999; Lama and Buchy, 2002; Sarin, 1998). Policy makers often overlook gender, class and castes and do not have understanding of individual relations that have constructed through gender, class and caste. Men and women form the poorer sections of the society, especially of the lower caste, have been excluded or prevented from participating in community forestry programme (Hobley, 1990; Lama and Buchy, 2002). Community forestry in Nepal fails to deliver its promise and instead dis-empowers those already marginalized. On the other hand, the local elite, rich families, upper caste group and leaders dominate decision-making process and as a result, get more benefit. This will in the long run alienate subaltern groups from mainstream development, hence threatening sustainable management of forest resources.

Although some researchers (e.g. Agarwal, 1986) have questioned the belief that foraging and fuelwood collection by the rural poor is primarily responsible for the shortages, the findings of these studies are ignored by development practitioners. It is commercial demand that have more frequently resulted in large-scale forest destruction (Shiva, 1999; Poffenberger 1995, 1996b; Correa, 1999). Earlier, in spite of govt. regulation, people would use the forests for firewood, manure and NTFPs. With the introduction of JFMP, people agree not to use (or to use specified areas) of the forest for these purposes, thereby restricting their use of the forest (Correa, 1999). It has been proved that community-based forest protection activities resulted in the rapid regeneration of degraded natural forests and confirmed the best prospects for sustainable forestry. Natural regrowth led to substantial increase in biomass productivity and enhanced availability of a range of important minor forest products. The capacity of degraded natural forests to rapidly regenerate and produce fodder, fuel, fibres and other valuable materials appears to have been instrumental in sustaining community protection activities (Poffenberger *et al.*, 1995, 1996b; Correa, 1999).

DATA SET AND FINDINGS

The study is based on secondary data relating to the study entitled "Role of NTFP in sustenance of JFM: a case-study", carried out by the Divisional Forest Officer (DFO) west Midnapore division under the office of the Principal Chief Conservator of Forest, Govt. of West Bengal in 2001. The study was conducted in four FPCs of Nayagram block, in Paschim Midnapore district. The enter block is very well forested with forest covering about 48 per cent of the total geographical area. As per last census, Nayagram block has a total population of 106490 of which 18.33 per cent is scheduled cast and 40.45 per cent is scheduled tribe. Sabar constitute one of the dominant ethnic groups. The main economic activity of the population is agriculture, the income from forest sources being the second most important source.

Almost entire forest was cut down during late '70s and early '80s of the last century mainly due to mass illicit felling by the contractors for the commercial needs. Since then, the forest resuscitated following the adaptation of joint forest management principles. The predominate species of the forest is sal with associates like mohua (madhuca indica), bahera (terminalia balerica), piasal (pterocarpus marsupium), asan (terminalia tomentosa), karam (adina cordifolia), pial (buchanania lanzan) etc. Most of the tree cops are of coppice origin. The study has conducted in following four FPCs:

1. **Ambisole FPC** – This forest protection committee is located in the Nayagram beat of Nayagram Range. The committee was registered in the year 1991, it's registration number being 392/ WM/FPC/224/NG-42. Total forest area under the forest protection committee is 130 hectares. There are 41 members in this FPC of which 40 belong to scheduled tribe and one to general caste. There are 23 Lodha members in this FPC. All the 40 scheduled tribe members of this FPC are landless.
2. **Bansiasole FPC** – This forest protection committee is also located in the Nayagram beat of Nayagram Range. The committee was registered in the year 1990, it's registration number being 330/ WM/FPC/45/NG-17. Total forest area protected by this FPC is 102 hectares. There are 60 members in this FPC, of which 34 belong to scheduled tribe and the rest 28 to general cast. There are 18 Lodha members in this FPC. All the 34-scheduled tribe member of this FPC is landless.
3. **Kasia FPC** – This forest protection committee is located in Chandabilla beat of Chandabilla Range. The committee was registered in the year 1991; its registration number being 378/ CB/WM. Total forest area protected by this FPC is 215.89 hectares. There are 100 members in this FPC, of which 24 belong to scheduled tribe and the rest to general caste. There are 11 Lodha members in this FPC. Fourty-one families are landless.
4. **Kadokata FPC** – This forest protection committee is also located in Chandabilla beat of Chandabilla Range. The committee was registered in the year 1995; its registration number being 409/ CB/WM. Total forest area protected in this FPC is 343.27 hectares. There are 24 members in this FPC. All of them are tribal. The forest protected by these FPCs shows typical characteristic of this area. Tree species in the forest were enumerated by laying out sample plots.

This empirical study might help us know whether the economic returns from forest – the share of net revenue from final harvesting of timber, income in the form of wages from forestry activities and income from NTFP in the forest – are significant for FPC member for the sustainability of JFM. But data in some cases are non-classified and non-tabulated. Despite these limitations, we have attempted to study their findings in simple proportions, averages and in some tabular forms for this paper. A mathematical model is also framed in the light of our empirical results (represents in the appendix).

As may be seen in table 1, the per capita annual money incomes from the share of revenue from timber are dissimilar across four FPCs studied. Since felling intervals are not same for all FPCs studied, area of felling coupe is not uniform and per capita forest of FPC members across the

Table 1
Income from the Share of Revenue from Timber from 1995-96 to 2000-01

Name of FPC	*No. of members*	*Share of revenue for the entire FPC (Rs.)* 1995-96	*1996-97*	*1997-98*	*1998-99*	*1999-00*	*2000-01*	*Income/ member/ year (Rs.)*
Ambisole	41	0	13373	0	79734	0	0	378 (1.48)
Bansiasole	60	0	15338	0	0	0	11984	76 (0.36)
Kasia	100	0	0	0	0	48084	0	80 (0.22)
Kadokata	24	0	0	0	0	0	80985	562 (1.37)
Total	225	0	28711	0	79734	48084	92969	185 (0.59)

Figures within brackets represent percentages in respect of income from timber out of total forestry income.

committees are dissimilar; also per capita notional income of a member per year is not equal. It is important to mention that each FPC, according to JFM agreement, becomes eligible to get a share of 25 per cent of total revenue earned from final harvesting of timber products. It is also observed that although per capita annual money income of a member of Ambisole and Kadokata FPCs is higher than that of total average. Out of total forestry income of FPCs, the share of revenue of final harvest of timber is 1.48 per cent in Ambisole FPC, 0.36 per cent in Bansiasole FPC, 0.22 per cent in Kasia FPC and 1 .37 per cent in Kadokata FPC. As the share of revenue from timber is very small source of income for all FPC members and they receive this income at some irregular annual intervals, this income, usually, does not meet up their immediate survival needs. It is important to mention that although the success of Arabari experience in JFM in Midnapore district, West Bengal is well known, however, ironically, in the same district, JFM of Arjuni mouza which started from 1991, failed to deliver results after the middle of 1994 as the members of FPC were only granted to have 25 per cent share of timber, without any other share of the forest resources which was insufficient to meet up the immediate survival needs of poor FPC members. It caused for large illicit felling, mainly, by the poor forest committees due to meeting up their immediate seasonal livelihood and food insecurity, which plagued the area and led to conditions of semi-starvation amongst the poor people.

In order to meet their immediate survival need, the poor FPC members are exclusively engaged as labour in forestry operations carried out in their respective jurisdiction. All such forestry operations, such as raising nursery, planting, tending, harvesting, entry point activities are labour intensive and serve as small scale of their income, particularly in the lean season (Table 2). Income from this source per member per year is computed by adding wage component of total amount spent over 5 years (1996-97 to 2000-01) in a particular FPC and dividing it, first, with strength of FPC members and then with number of years (five years in this case). This

Table 2
Income as Wage from Forestry Works from 1996-97 to 2000-01

Name of FPC	*No. of members*	*Amount spent in the FPC in this five years (Rs.)*	*Income/member/ year (Rs.)*
Ambisole	41	251629.33	1227.64 (4.80)
Bansiasole	60	355459.40	1184.86 (5.67)
Kasia	100	432531.18	865.06 (2.34)
Kadokata	24	539749.97	4497.91 (10.98)
Total	225	1579369.88	1403.88 (4.53)

Figures within brackets represent percentages in respect of income as wage from forestry work out of total forestry income.

source of income, too, is variable across FPCs and depends on the scope of forestry operation in a particular committee area. Income as wages from forestry works constitute 4.80 per cent in Ambisole FPC, 5.67 per cent in Bansiasole FPC, 2.34 per cent in Kasia FPC and 10.98 per cent in Kadokata FPC out of their total income from the forestry sector. Although this source is small, it is one of their sources of survival during lean season (June to September).

In addition to the maintenance of regular consumption needs of the local FPC members, NTFP is the main source of money income for all FPC members under our study (Tables 3a, 3b, 3c and 3d). Some common characteristics may be discerned in this regard. First, the money income generated from the share of revenue from timber and income as wage from forestry works are insignificant in relation to NTFPs for all FPCs studied. The contribution of NTFPs' income of Ambisole, Bansiasole, Kasia and Kadokata FPCs are 93.72 per cent, 93.98 per cent, 97.44 per cent and 87.65 per cent respectively out of the total money income (from all sources) of the respective FPCs. Secondly, sal leaves (shorea robusta) are the main source of money income of NTFPs for all FPC members. More than 35 per cent of total NTFPs' incomes for all FPC members come from the sale of sal leaves to the local market. The period of availability of sal leaves is the highest (10 months in a year) of all NTFPs. The money income received from other NTFPs is different for different FPCs. Thirdly, out of 36 types of NTFPs, 10 types[11] are common to all FPCs studied. The contribution of these common NTFPs' income of Ambisole, Bansiasole, Kasia and Kadokata FPCs works out to 62.58 per cent, 70.18 per cent, 62.20 per cent and 78.64 per cent respectively out of total NTFPs' income of the respective FPCs. Fourthly, the period of availability for the same NTFP does not differ among FPC members. Moreover, the local rate (Rs. per unit) of the same NTFP does not differ among FPC members. The members of each FPC dispose of the same NTFP at a fixed price. The amount of output the members of the FPCs decides to sale seems to have no effect on the local market of the product.

Table 3a

Period of Collection, Sale Value, Collection Intensity and Annual Per Capita Income from NTFPs of Ambisole FPC

NTFPs	*Period of collection*	*Local rate (Rs.)*	*Average daily collection (per member)*	*Average annual income (per member/Rs.)*
Sal (*Shorea robusta*) leaves	10 months	14 per thousand	2000 nos.	8400 (35.04)
Pial (*Buchanania lanzan*) fruit	21 days (Apr - May)	20 per kg.	1 kg	420 (1.75)
Mohua (*Madhuca indica*)	30 days (Apr - May)	8 per	90 kg	720 (3.00)
Mohua flower	45 days	8 per, kg	2.5 kg	900 (3.76)
Bahera (*Terminalia balerica*) fruit	30 days (Mar - Apr)	3 per kg	5 kg	450 (1.88)
Kurchi (*Holarrhena antidysenterica*) fruit	30 days (Feb - Mar)	22 per kg	1 kg	660 (2.76)
Haritaki (*Terminalia chebula*) fruit	15 days	10 per kg	3 kg	450 (1.88)
Dudhi lata (*Oxystelma esculata*)	30 days	4 per hundred	400 nos.	480 (2.01)
Kurkura (*Lycoperdon sp*) mushroom	30 days (June - July)	10 per kg	4 kg	1200 (5.00)
Karam mushroom	30 days	20 per kg	1 kg	600 (2.50)
Ghora insect	30 days	130 per kg	0.3 kg	1170 (4.88)
Rahara (*Soymida febrifuja*) fruit	20 days (May - June)	6 per hundred	300 nos.	360 (1.50)
Sidha (*Lagerstroemia parviflora*) fruit	15 days	3 per kg	2 kg	90 (0.38)
Anantamul (*Hemidesmus indicus*)	15 days	20 per kg	2 kg	600 (2.50)
Satamul (*Asperagus racemosus*)	15 days	6 per kg	2 kg	180 (0.75)
Chun alu (*Dioscorea deltoidea*)	120 days	6 per kg	2 kg	1440 (6.01)
Kalmegh	15 days	7 per kg	6 kg	630 (2.63)
Paan alu	120 days	6 per kg	1 kg	720 (3.00)
Kanta alu (*Dioscorea pentaphylla*)	120 days	9 per kg	2.5 kg	2700 (11.26)
Sal resin	60 days	60 per kg	0.5 kg	1800 (7.51)
Total				23970* (100)

Figures within brackets represent percentages of income out of total income from NTFPs.

*The percentage of NTFPs' income for Ambisole FPC members works out to 93.72 out of their forestry income.

Table 3b

Period of Collection, Sale Value, Collection Intensity and Annual Per Capita Income from NTFPs of Bansiasole FPC

NTFPs	*Period of collection*	*Local rate (Rs.)*	*Average daily collection (per member)*	*Average annual income (per member/Rs.)*
Sal (*Shorea robusta*) leaves	10 months	14 per thousand	2000 nos.	8400 (42.67)
Pial (*Buchanania lanzan*) fruit	21 days (Apr - May)	20 per kg.	0.5 kg	210 (1.07)
Mohua (*Madhuca indica*)	30 days (Apr - May)	8 per	30 kg	240 (1.22)
Mohua flower	45 days	8 per kg	0.5 kg	180 (0.91)
Bahera (*Terminalia balerica*) fruit	30 days (Mar - Apr)	3 per kg	10 kg	900 (4.57)
Kurchi (*Holarrhena antidysenterica*) fruit	30 days (Feb - Mar)	22 per kg	1 kg	660 (3.35)
Haritaki (*Terminalia chebula*) fruit	15 days	10 per kg	2 kg	300 (1.53)
Dudhi lata (*Oxystelma esculata*)	30 days	4 per hundred	500 nos.	600 (3.05)
Kurkura (*Lycoperdon sp*) mushroom	30 days (June - July)	10 per kg	3.5 kg	1050 (5.34)
Karam mushroom	30 days	20 per kg	2 kg	1200 (6.09)
Ghora insect	30 days	130 per kg	0.25 kg	975 (4.95)
Ban karala (Urea lobata)	30 days	25 per kg	0.1 kg	75 (0.38)
Anantamul (*Hemidesmus indicus*)	15 days	20 per kg	1 kg	300 (1.52)
Satamul (*Asperagus racemosus*)	15 days	6 per kg	3 kg	270 (1.37)
Chun alu (*Dioscorea deltoidea*)	120 days	6 per kg	1 kg	720 (3.66)
Kalmegh	15 days	7 per kg	5 kg	525 (2.67)
Paan alu	120 days	6 per kg	0.5 kg	360 (1.83)
Kanta alu (*Dioscorea pentaphylla*)	120 days	9 per kg	1.5 kg	1620 (8.23)
Sal resin	60 days	60 per kg	0.25 kg	900 (4.57)
Ban kundri (Coccinia indica)	15 days (May)	20 per kg	10 kg	200 (1.02)
Total				19685* (100)

Figures within brackets represent percentages of income out of total income from NTFPs.
*The percentage of NTFPs' income for Bansiasole FPC members works out to 93.98 out of their forestry income.

Table 3c

Period of Collection, Sale Value, Collection Intensity and Annual Per Capita Income from NTFPs of Kasia FPC

NTFPs	*Period of collection*	*Local rate (Rs.)*	*Average daily collection (per member)*	*Average annual income (per member/Rs.)*
Sal (*Shorea robusta*) leaves	10 months	14 per thousand	3500 nos.	14700 (40.89)
Pial (*Buchanania lanzan*) fruit	21 days (Apr - May)	20 per kg.	2.5 kg	1050 (2.92)
Mohua (*Madhuca indica*) fruit	30 days (Apr- May)	8 per kg	120 kg	960 (2.67)
Mohua flower	45 days	8 per	3 kg	1080 (3.00)
Kurchi (*Holarrhena antidysenterica*) fruit	30 days (Feb - Mar)	22 per kg	2 kg	1320 (3.67)
Haritaki (*Terminalia chebula*) fruit	15 days	10 per kg	2 kg	300 (0.83)
Dudhi lata (*Oxystelma esculata*)	30 days	4 per hundred	150 nos.	180 (0.50)
Kurkura (*Lycoperdon sp*) mushroom	30 days (June - July)	10 per kg	2 kg	600 (1.67)
Karam mushroom	30 days	20 per kg	2 kg	1200 (3.34)
Ghora insect	30 days	130 per kg	0.25 kg	975 (2.71)
Bhurru (*Gardenia gummifera*) fruit	20 days	4 per kg	5kg	400 (1.11)
Atari (*Combretum decundrum*) fruit	15 days	2 per kg	20 kg	600 (1.67)
Anantamul (*Hemi-desmus indicus*)	15 days	20 per kg	2 kg	600 (1.67)
Asan (*Terminalia tomentosa*) gum	120 days	10 per kg	1 kg	1200 (3.34)
Chun alu (*Dioscorea deltoidea*)	120 days	6 per kg	1.5 kg	1080 (3.05)
Bahera gum	120 days	10 per kg	2 kg	2400 (6.67)
Paan alu	120 days	6 per kg	1 kg	720 (2.07)
Kanta alu (*Dioscorea pentaphylla*)	120 days	9 per kg	2 kg	2160 (6.01)
Sal resin	60 days	60 per kg	0.25 kg	900 (2.50)
Ban kundri (*Coccinia indica*)	15 days (May)	20 per kg	20 kg	400 (1.11)
Pog mushroom	60 days (July-Aug)	8 per kg	1.5 kg	720 (2.00)
Jara kalai (*Glycine hispida*)	75 days (June-Aug)	30 per kg	80 kg	2400 (6.67)
Total				35945* (100)

Figures within brackets represent percentages of income out of total income from NTFPs.

* The percentage of NTFPs' income for Kasia FPC members works out to 97.44 out of their forestry income.

Table 3d

Period of Collection, Sale Value, Collection Intensity and Annual Per Capita Income from NTFPs of Kadokata FPC

NTFPs	*Period of collection*	*Local rate (Rs.)*	*Average daily collection (per member)*	*Average annual income (per member/Rs.)*
Sal (*Shorea robusta*) leaves	10 months	14 per thousand	5000 nos.	21000 (58.48)
Pial (*Buchanania lanzan*) fruit	21 days (Apr- May)	20 per kg	5 kg	2100 (5.85)
Mohua (*Madhuca indica*) fruit	30 days (Apr- May)	8 per kg	40 kg	320 (0.89)
Mohua flower	45 days	8 per kg	2 kg	720 (2.01)
Kurchi (*Holarrhena antidysenterica*) fruit	30 days (Feb - Mar)	22 per kg	0.5 kg	330 (0.92)
Haritaki (*Terminalia chebula*) fruit	15 days	10 per kg	3.5 kg	525 (1.46)
Dudhi lata (*Oxystelma esculata*)	30 days	4 per hundred	50 nos.	60 (0.17)
Kurkura (*Lycoperdon sp*) mushroom	30 days (June - July)	10 per kg	5 kg	1500 (4.18)
Karam mushroom	30 days	20 per kg	1.5 kg	900 (2.51)
Ghora insect	30 days	130 per kg	0.2 kg	780 (2.17)
Bahera (*Terminalia balerica*) fruit	30 days (Mar-Apr)	3 per kg	3.5kg	315 (0.88)
Atari (*Combretum decundrum*) jhanti	60 days	10 per basket	0.5 pon.	3000 (8.35)
Asan (*Terminalia tomentosa*) fruit	30 days	2 per kg	10 kg	600 (1.67)
Asan (*Terminalia tomentosa*) gum	120 days	10 per kg	0.5 kg	600 (1.67)
Bahera gum	120 days	10 per kg	1.5 kg	1800 (5.01)
Ban kundri (*Coccinia indica*)	15 days (May)	20 per kg	20 kg	400 (1.11)
Pog mushroom	60 days (July-Aug)	8 per kg	2 kg	960 (2.67)
Total				35910* (100)

Figures within brackets represent percentages of income out of total income from NTFPs.

* The percentage of NTFPs' income for Kadokata FPC members works out to 87.65 out of their forestry income.

This study, however suggests that the relative importance of NTFPs in forest-based economics is supreme. Consequently, it plays the major role in sustenance of joint forest management and to this end the FPC members are expected to optimize the production from timber to NTFPs (discussed in the appendix with a mathematical model). Although, the sample size (four FPCs) is too small to make a generalization, it may be said that in predominantly tribal FPCs with a good cover of sal forest, NTFPs provide the main source of forestry income.

CONCLUSION

The historical context in which JFM has emerged as the latest in a history of policy changes in India has been grounded since long past by the community resistance movement of local poor communities against the old custodian system of forest management. In many parts of India where forests are an essential component of the local livelihood support system, a community-based rights regime with build-in safeguard for access and the livelihood of forest dependent people have been grounded historically and that have provided a solution to problems of sustainable forest management as well as sustainable local livelihood maintaining bio-diversity, ecological balance and environmental stability. This study also suggests that the present JFM model needs to be remodeled in terms of livelihood benefits of the poor forest communities which are immediate and less commercial and in line with this, the new policy management of both degraded and non-degraded forests are to be re-oriented from timber production to optimizing the production of NTFPs for strengthing the livelihood of local communities as well as the sustainability of community forest management with environmental stability. Any JFM, which does not recognize significance of sustaining livelihood – food security – at the local level, have a doubtful future.

APPENDIX

MODEL: The solution to optimal resource exploitation problems of community forestry relating to the sustainability of JFM system may be studied by the form of present-value Hamiltonian function. This form assumes that simple replication of past decision with any other choice (there is none, in fact) to be made at some later date. Each decision imposes an externality on the future (Silberberg, 1990; p. 14-15). The solutions to this problem are termed 'open-loop control'. They represent once-and-for-all solutions to dynamic optimization problems, which are as if the firm decides in the initial period what to do for all future periods and stick to that plan. The following assumptions are taken in this regard.

Firstly, the growth function of NTFPs is a continuous time-logistic function

$$v(t) = g(v(t), u(t), t)$$

where $v(t)$ is the stock and $u(t)$ is the rate of harvest. The logistic growth function is widely used in empirical analysis (Hanley *et al.*, 1997: chapter 10). The growth of NTFPs may continue for some years although ultimately the volume of NTFPs reaches at its peak and then declines as the NTFPs decay and eventually die. It has the characteristics that at low stock rate of growth is low, it peaks at a particular time and then declines towards zero.

Secondly, the total income (nominal and/or physical) of NTFPs is shared equally among the members of local forest communities involved in JFM program. Government does not take any share of income from NTFPs.

Thirdly, the members of FPC in JFM sell the NTFPs in a competitive market at a fixed price (discussed in the text). pu is the net revenue function and u is the control variable.

Finally, both harvesting and planting of non-timber trees are costless.

JFM committee aims to maximize its profit subject to the stock over an infinite time interval. Mathematically the problem is

$$\text{Max}_u \int_o^\infty pu\, e^{-rt}\, dt \tag{1}$$

subject to $\dot{v} = g(v, u, t),\ v(o) = v_o$

where u is the harvest rate at time t (NTFPs are usually harvested every year). v is the stock, pu is the net revenue function and $g(.)$ is a logistic growth function. $x(o)$ is the initial stock [Endpoint conditions vary. Typically the initial stock is fixed, although the final stock may not be (Silberberg, 1990; p. 618)]. r is the discount rate. The variables v, u and λ all have time subscripts. The growth function, $g(v, u, t) = av + bv^2 - u, a > 0;$ $b < 0$.

The necessary conditions can be represented by introducing Hamiltonian function, $H(.)$

$$H(v, u, \lambda, t) = pue^{-rt} + \lambda g(v, u, t)$$

This function is equal to the profit plus the change in the stock valued by its shadow price. The Hamiltonian allows a convenient representation of the necessary conditions which compromise the maximum priciple: first differentiation $H(.)$ with respect to u and setting equal to zero.

$$\frac{\delta H}{\delta u} = pe^{-rt} + \lambda g_u = 0 \tag{2}$$

where the Hamiltonian is maximised with respect to u, and the costate condition is

$$\frac{\delta H}{\delta u} = \lambda = - \lambda g_v \tag{3}$$

The costate variable is not of direct interest, so (2) and (3) is used to eliminate it. In the same series of manipulations, the time variable is also eliminated so that results can be expressed independently of time.

First writing (2) explicitly for the logistic growth function

$$pe^{-rt} - \lambda = 0 \tag{4}$$

and note that this implies that harvest continues until the marginal profit (here average profit) equal the marginal value of stock, λ.

Differentiating (4) with respect to time

$$\lambda = -r\, pe^{-rt} \tag{5}$$

When the specific functional form (logistic growth function) is included, (3) can be written as

$$\lambda = -\lambda\,(a + 2bv) \tag{6}$$

The procedure now is to eliminate the costate variable and define a steady-state solution.

To this end, equate (5) with the (6) and eliminate λ using (4) to define a steady-state equilibrium

$$r = a + 2bv = g_v \tag{7}$$

where own rate of return on the stock (g_v) is equal to the discount rate (r).

The term g_v is of great importance: it indicates how the rate of stock growth changes with respect to the stock and it, therefore, represents the return on relating the marginal unit of stock. Thus gv is the own rate of return on the stock.

The system of differential equations is complete by the growth equation

$$v = \mathrm{g(u,v)} = av + bv^2 - u \tag{8}$$

This is the required result, it represents the problem as two autonomous differential equations (7) and (8) which are independent of time. Although there is no guarantee that these equations can be solved analytically, they allow us to assess the optimal harvest rate of NTFPs given initial conditions (v_0, u_0) and a steady-state solution which may be viewed as the comparative static solution to the dynamic problem where the stock and the harvest rate are constant and no incentives exist to adjust them, to the problem (The equilibrium solution in an imperfect market is identical to that of the competitive problem [7]).

The necessary and sufficient condition for the solution to (1) can be achieved by checking that second order derivative of the maximized Hamiltonian with respect to v is non-positive

$$\frac{\delta^2 H}{\delta v^2} = \lambda\, 2b$$

As $\lambda \geq 0$; that is, shadow price of stock is non-negative and $b < 0$, the condition for concavity is satisfied and a solution which satisfies the maximum principle conditions is both necessary and sufficient (Hanley *et al.*, 1997; p. 200).

If we, however, relax the last assumption and introduce cost function (cost per unit harvesting is declining function of the stock), the problem becomes

$$\text{Max}_u \int_o^\infty (pu - c(v)\, q)\, e^{-rt}\, dt$$

subject to $v = g(v, u, t),\ x(o) = x_o$

and the result after manipulation will be

$$g' v) = \frac{c'(v)g(v)}{(p - c)v))} = r \tag{9}$$

From (9) the rate of return on holding the marginal unit of stock can be decomposed into two parts: the return from increased stock growth; $g'(v)$ and the return from reduce cost. This implies that the optimum level of stock is less in the presence of costs than would be case for zero costs.

ACKNOWLEDGRMENTS

The author would like to thank Mr. Debal Roy, Divisional Forest Officer, west Midnapore division and Mr. Abhijit Basu Roy Choudhury, Chief Conservator of Forest Research, Department of Forest, Government of West Bengal, for providing data of this paper. Special thanks to Mr. Nihar Ranjan Chakroborty, DFO (Statistics) and Mr.Gopal De, Assistant Statistician, Department of Forest, Government of West Bengal, for helpful comments.

NOTES

1. See Duyker, Tribal Guerrillas, p. 28.
2. For detailed information, see L. S. S. O'Malley, Bengal District Gazetter: Bankura (Calcutta 1911), W. W. Hunter, Statistical Account of Bengal: Vol-III, (London, 1876).
3. The links in tribal belief between the health of the forest, fertility and prosperity are clear in the following lines from this Baha festival song.

   When the sal trees are in leaf,
   On the mountain,
   How lovely they look,
   Wealth in the house....

   See W. G. Archer, The Hill of Flutes, Love and Poetry in Tribal India: A Portrait of the Santals (London, 1974), p. 237, for and extensive discussion.
4. See S. Dasgupta, Adivasi Policies in Midnapore.
5. The British adopted the Bengali term 'chaur' meaning an outlandish or wild person, to refer to the tribal and low-caste people of the area.
6. For detailed information, see Duyker,Tribal Guerrillas, p. 35.

7. See Dutta (Calcutta, 1940) the Santal Insurrection of 1855-1857, p. 26.
8. For discussion, see Duyker,Tribal Guerrillas, p. 35.
9. It is likely that similar community concerns over environmental degradation in other parts of India could provide effective support for joint management programmes if initiated by state forest departments.
10. The common factors are (i) perceived benefits from co-operating; (ii) clearly defined rights and boundaries for resources; (iii) knowledge about the state of the resources; (iv) small size of the user group; (v) low degree of heterogeneity of the user group; (vi) long-term, multi-layered interaction among the community; (vii) simple rules and adaptable management regimes; (viii) graduated sanctions as punishment; (ix) ease of monitoring, and accountability; (x) conflict resolution mechanisms and the role of leadership; (xi) influences from the wider political economy.
11. The common NTFPs are sal leaves, pial fruit, mohua fruit, mohua flower, kurchi fruit, haritaki fruit, dudhi lata, kurkura mushroom, karan mushroom, ghora insect.

REFERENCES

Agarwal, B. (1997), 'Environmental Action, Gender Equity and Women's Participation', *Development and Change,* 28(1): p. 1-44.

'———' (1986), 'Cold Hearths and Barren Slopes*: The Wood fuel Crisis in the Third World'*, New Delhi: Allied Publications.

Agarwal, C. and S. Saigal (1996), 'Joint Forest Management in India: A Brief review', SPWD working paper, New Delhi, 16 May.

Archer, W. G (1974), 'The Hill of Flutes, Love and Poetry in Tribal India', *A Portait of the Santals,* London. Baden-Powell, B. H. (1892), *Law system of British India:* Vol.-I, Oxford.

Baland, J. M. and J. P. Platteau (1996), *Halting Degradation of Natural Resources: Is there a Role for Rural Communities?* Oxford: Clarendon Press.

Bannerjee, U. (1989), 'Participatory Forest Management in West Bengal', In K. C. Malhotra and Mark Poffenberger, eds, *Forest Regeneration through Community Participation* (West Bengal Forest Department): p. 4.

Berkes, F. (ed.) 1989, 'Common Property Resources'. *Ecology and Community Based Sustainable Development,* London: Belhaven Press.

Bromley, D. W. (1992), *Making the Commons Work: Theory, Practice, and Policy (ed.),* Institute for Contemporary Studies Press: San Francisco.

Chatterjee, G. (1986), *'Report from the Resident of Midnapore, 6 February 1773'*, The Forerunner of India's Freedom Struggle, Delhi: p. 38.

Chain, A. C. (1984), *Fundamental Methods of Mathematical Economics,* 3rd edn., New York: McGraw-Hill.

'———' (1992), *Elements of Dynamic Optimization,* New York: McGraw-Hill.

Ciriacy-Wantrup, S. V. and R. C. Bishop (1975), 'Common Property as a Concept in Natural Resource Policy', *Natural Resources Journal* 15: 713-27.

Correa, M. (1999), 'The Need for Emancipatory Research: Experiences from JFPM in Uttara Kannada', In R. Jeffery and N. Sundar, eds, *A New Moral Economy for India's Forests?* New Delhi: Sage Publications.

Crabbe, P. J. and N. V. Long (1989), 'Optional Forest Rotation Under Monopoly and Competition', *Journal of Environmental Economics and Management.*

Duyker, E. (1987), *Tribal Guerrillas: The Santals of West Bengal and the Naxalite Movement,* Delhi.

Hanley, N. J. F. Shogren and B. White (1997), *Environmental Economics – in Theory and Practice,* New Delhi: Macmillan.

Hanley, N., A. Munro, D. Jamieson and D. Ghosh (1991), *Environmental Economics and Sustainable Development in Nature Conservation,* Peterborough: NCC.

Hobley, M. (1996), *'Participatory Forestry: Process of Change in India and Nepal', Rural Development Forestry Study Guide* 3, Overseas Development Institute, London.

Hunter, W. W. (1876), *Satistical Account of Bengal,* Vol.-III: p. 18, London.

Lama, A. and M. Buchy (2002), 'Gender, Class, Caste and Participation: The Case of Community Forestry in Nepal', *Indian Journal of Gender Studies,* 9:1, New Delhi: Sage Publications.

Locke, C. (1999), 'Women's Representation and Roles in Gender Policy in JFM', In R. Jeffery and N. Sundar, eds, *A New Moral Economy for India's Forest - Discourses of Community and Participation,* New Delhi:Sage Publications.

Martin, F. (1992), 'Common Pool Resources and Collective Action: A Bibliography', *Paper Presented to the Workshop on Political Theory and Political Analysis,* Bloomington, Indiana.

McAlpin, M. C. (1909), *Report on the Condition of the Sonthals in theDistricts of Birbhum Bankura, Midnapore and North Balasore,* Calcutta.

McKean, M. A. (1992), 'Management of Traditional Common Lands (Iriachi) in Japan', In D. W. Bromley, eds, *Making the commons work: Theory, Practice and Policy,* San Francisco: ICS Press.

Mukherjee, N. (1995), 'Forest Management and Survival Needs: Community Experience in West Bengal', Economic *and Politically Weekly,* 30(49): 3130-132.

Mukherjee, S. (1975), 'A chapter on the tribal sources of Bengal history in the Muslim period', *Bulletin of the Cultural Research Institute,* Delhi: 11, 1-2: p. 20-4.

Naik, G. (1995), 'Determinants of People's Participation in Joint Forest Management (mimeo)', *Indian Institute of Management,* Ahmedabad.

'———' (1997), 'Forest Management: Factors Influencing Household Participation', *Economic and Political Weakly,* November 29.

O'Malley, L. S. S. (1911), *'Bengal District Gazetteer: Bankura'*, Calcutta.

Ostrom, E. (1990), *Governing the commons: The Evolution of Institutions for Collective Action*, Cambridge University Press.

Poffenberger, M. (1995), 'The Resurgence of Community Forest Management in the Jungle Mahals of West Bengal', *Nature, Culture, Imperialism - Essays on the Environmental History of South Asia*, David Arnold and Ram Chandra Guha, eds, Delhi: Oxford University Press.

'———' (1996a), 'The Stuggle for Forest Control in the Jungle Mahals of West Bengal, 1750-1990', In M. Poffenberger and B. McGean, eds, *Village Voices, Forest Choices: JFM in India, Delhi:* Oxford University Press.

Poffenberger, M., B. McGean and A.Khare (1996b), 'Communities Sustaining India's Forest in the Twenty-first Century', *Village Voices, Forest Choices: JFM in India*, Delhi: Oxford university Press.

Roy, D. (2001), 'Role of NTFP in Sustenance of Joint Forest Management – a case study', Office of the Principal Chief Conservator of Forest, Govt. of West Bengal.

Sarin, M. (1998), 'Who is Gaining? Who is Losing?', *Gender and Equity Concerns in Joint Forest Management*, SPWD.

Sarker, D. N. and N. Das (2002), 'Women's Participation in Forestry – Some Theoretical Issues', *Economic and Political Weakly*, October 26: p. 4407-12.

Saxena, N. C. and M. Sarin (1999), 'The Western Ghats Forestry and Environmental Project in Karnatka: A Preliminary Assessment', In R. Jeffery and N. Sundar, eds, *A New Moral Economy for India's Forests*? New Delhi: Sages Publications.

Sherwill, W. S. (1885), *'Geographical and Statistical Report of the District of Beerbhoom'*, Calcutta: p. 31.

Singh, C. (1995),'Forests, Pastoralists and Agrarian Society in Mughal India', In D. Arnold and R. C. Guha, eds, *Nature, Culture. Imperialism: Essays on the Environmental History of South Asia*, Delhi: Oxford University Press.

Singh, N. M. (2001), 'Women and Community Forests in Orissa: Rights and Management', *Indian Journal of Gender Studies*, 8:2, Sage Publication, New Delhi.

Solow, R. M. (1974), 'The Economics of Resources or the Resousces of Economics', *American Economic Review*, 64: 1-14.

Sundar, N., A. Mishra and N. Peter (1996),'Defending the Dalki Forest: Joint Forest Management in Lapanga', *Economic and Political Weakly* 31 (45 & 46): 3021-25.

Vira, B. (1999), 'Implementing JFM in the field: Towards an Understanding of the Community – Bureaucracy Interface', Ch.-11, In R. Jeffery and N. Sundar, eds, *A New Moral Economy for India's Forest*? New Delhi: Sage Publications.

28

INSTITUTIONAL CHANGE AND COMMON PROPERTY RESOURCE MANAGEMENT IN SHIVALIKS, JAMMU AND KASHMIR[#]

Falendra K Sudan

INTRODUCTION

In the recent past, experiences have been gained on institutional change in irrigated agriculture, community and social forestry, horticultural development, fisheries development, rural development etc, and have influenced the development policies even though they may not have yet reached the implementation and operational level. However, not many experiences have been available or documented, which deal with institutional changes to multi-sectoral, multi-objective programmes such as integrated watershed management, particularly in mountainous and rainfed areas. This is because; they are very complex programmes often dealing with conflicting objectives e.g. development and environment, maximization of economic benefits and conservation of resources, market economy and self-help/food security. However, attempts are being made to facilitate the institutional change in this area also (Sharma, 1995).

Under Integrated Watershed Development Project (IWDP), Hills-II, Jammu and Kashmir institutional reforms have been implemented for watershed development and protection in various sub-sectors such as provisioning of agricultural services, forestry, water resources, etc. The local level stakeholders have been entrusted with the task of actual management of natural resources to make more productive use of those resources. It has been emphasized that the users' participation in resource management would ensure proper maintenance and sustainability of the

This paper is a part of the research project titled "Participatory Approaches and Environmental and Economic Impact: With Special Reference to IWDP, Hills-II, Jammu and Kashmir, India", completed in May 2003 and sponsored by Ministry of Environment and Forests, New Delhi and World Bank funded "Environmental Management Capacity Building Technical Assistance (EMCaB) Project".

assets created over the project period and would facilitate smooth withdrawal and improve the performance of the project. The existing natural resources in the *Shivaliks* of Jammu and Kashmir share three essential features, which support institutional change in natural resource management. First, the ownership of forests, water resources and pastures and grazing lands is vested with the state, which does not have the capacity to regulate their access and use. Second, the individual resource management is difficult due to lack of defined boundaries in case of water resources and pastures and grazing lands, and at the same time the cost of supervision is supposed to be high. Third, the most natural resources have multiple users and uses; therefore private management ignores the equity aspects of natural resource management. These issues are now widely accepted and user participation in natural resource management has been widely encouraged under the institutional reforms. It has also been argued that common property resources (CPRs) are more effectively and efficiently managed through users' participation than state's regulation and control. The user groups have a direct stake in the future of the resource, which they are managing. They have better and easy access to information about local needs, so that they can take prompt action as per changed situation. The stability and robustness are essential qualities for user groups for successful management of natural resources (Wade, 1988; Ostrom, 1990). Keeping above in view, an attempt has been made to examine common property resource (CPR) management and the cost and benefit sharing mechanisms in Shivaliks, Jammu and Kashmir in the context of IWDP, Hills-II.

STUDY AREA AND METHODOLOGY

In order to prevent and reverse the degradation process of the *Shivaliks and Karewas*, Integrated Watershed Development Project (IWDP), Hills-I was undertaken in 1990. Seeing the achievement of physical targets, the World Bank decided to further extend the programme in the form of IWDP (Hills-II) in 1999 for five years to other rainfed areas of the state viz. Akhnoor and Ramnagar in *Shivaliks* (Jammu region) and Rajwar and Rambiyara in *Karewas* (Kashmir Valley). The main objective of the IWDP (Hills-II) is to restore the productive potential of the *Shivaliks* and *Karewas* by evolving participatory approaches. The focus is on improving *in-situ* moisture conservation, water harvesting and land management practices. An associated objective of the project is to assist with the institutional development and consolidate the progress already made under IWDP (Hills-I) (World Bank, 1999).

The present study is confined to two sub-watersheds of Akhnoor (Jammu district) and Ramnagar (Udhampur district). While Ramnagar is in the inner *Shivaliks*, Akhnoor is on its outer part. Ramnagar sub-watershed is the catchment area of Ramnagarwali *Khad* (gorge) in the middle catchment of river Tawi. It has an area of 32,630 Ha. and is subdivided into 39 micro-watersheds. Akhnoor sub-watershed with an area of 42350 Ha. is sub divided into 37 micro-watersheds. It covers about a dozen rivulets (*nallahs*) and a large number of small nallahs originating from *Kalidhar* ridge and draining into Chenab river on the western portion of *Kalidhar* ridge, drains join Manawar Tawi which in turn also meets the Chenab river. The *Shivaliks*, locally known, as *'Kandi'* comprise of piedmont deposits made of boulder and pebbles, gravel and sand with minor clays mixed in varying proportions. Due to lack of irrigation, subsistence rainfed agriculture is the prevalent production system is selected sub-watersheds. In terms of physiographic features, there may not be 100 per cent similarity between the two sub-watersheds but, certainly, in term of accessibility, natural resource endowment, infrastructure development and general index of socio-economic awareness, two sub-watersheds are comparable.

The primary data has been collected using well-structured and pre-tested questionnaires, participatory rural appraisal (PRA) techniques and group meetings. Whenever possible and required, secondary data sources have also been utilized. A total of six village Development Committees (VDCs), three each from selected sub-watersheds of Ramnagar and Akhnoor have been selected for collection of primary data and information. The stratified sampling technique has been used to select villages where VDCs have been created. During the stratification, care has been taken to include both "forested watersheds' villages and agricultural watersheds' villages" in order to makes a comparative study. The content analysis technique has been used to analyze the data and information qualitatively and quantitatively (using descriptive statistics).

INSTITUTIONS AND NATURAL RESOURCE MANAGEMENT

The weak community's intervention and erosion of grassroots mechanisms to protect and enhance rural community's stakes constitute the vital reason behind environmental resource degradation irrespective of economic status of the communities. The institutional approach in natural resource management is largely ignored in watershed areas, which apparently explain the link between environmental resource degradation and rural poverty (Jodha, 1986). During the recent past, some kinds of changes in institutional structure are seen to be playing a useful role. Small communities, often with non-government and/or government support are seen to come together with the express intention of preserving the environmental resources linked with consumption patterns and livelihoods (Blomquist, 1992; Wade, 1988; Tang, 1992; Uphoff, 1993). Such change in the institutional framework amount to an alternation in the nature of property rights. The newly emergent institutions lay down the nature of user rights and obligations. They aim at laying down a codified set of rules, which by convention or law mediates the nature of the relationship between resources and people. One of the outcomes of such new organizational form is an increase in the capacity of the environment to improve rural livelihoods.

A local level "institution creating" focus begins with examination of rights to access, use and ownership of resources by individuals, households and other social groups. It confronts the existing set of social norms and practices and questions whether they are appropriate in the context of efficient and sustainable resource management. If the need for new boundaries of access, or norms of behaviour arises, the micro-initiative had to put these in place and the following steps may be necessary in this process: (i) Consensus building with respect to the need for a new dispension, (ii) The setting out of detailed rules for working together on specific asset or income creation activities need to be worked out, (iii) The putting in place of a system of responsibility for repair and maintenance of assets created, (iv) The sharing of output, its time and the manner in which it accrues need to be spelt out as well, (v) The putting in place of a mechanism for changes in rules and norms of behaviour (Bromley, 1989; Kolavalli, 1997; Quiggin, 1993; Chopra and Kadekodi, 1998). The local issues led intervention is more flexible in the initial stages. It draw its strength from this ability to improvise rules and conventions of behaviour and to build on traditional institutions to set up innovative models of management, in particular in the context of natural resources. However, expansion may bring it in conflict with other institutions that pervade the economy at the macro level. Examples are legal bottlenecks and/or market generated constraints. There is greater need for decentralization and state involvement to ensure that the seat of local decision-making is not usurped by the politically and economically powerful (Dasgupta, 1997).

Institutionally sustainable rehabilitation and management of watershed development project is possible only where stakeholders/ beneficiaries act jointly to manage the resources. Watershed development is essentially a resource-base approach to environmental protection and livelihood enhancement. Unless adequate safeguards can be built in, the danger is that, as the commons become more productive, better-off farmers are tempted to take control of them and customary access rights of the poor are denied (Farrington *et al.*, 1999). Watershed development project is being affected by three broad groups of factors: How the project is designed and implemented, how the project is organised and what external agencies are operating at watershed level. Community awareness, knowledge, support and public participation in the decisions that affect the environment and economic outcomes are equally important. This is, best secured by creating centralized systems of regulation, use and management of resources upon which local communities depend and giving these communities an effective control over the use of these resources (WCED, 1987). To the extent over exploitation of natural resources is due to lack of inappropriate structure of property rights, creating and enforcing appropriate property rights in common property resources could be an instrument of averting their tragedy and improving the environmental and economic gains (Singh, 1997). India's experience with various watershed development projects shows that the participatory approach could ensure sustainable use of renewable common pool natural resources (Singh, 1995).

COMMON PROPERTY RESOURCE MANAGEMENT

During the recent past, a policy decision was made by the Government, which made it possible to hand over rights of usufruct on community lands to groups such as village *panchayats*, or to individuals on request, for appropriate land use. It was observed that this has generally not occurred. The farmers are apparently not willing to use these lands unless they are given land use titles by the revenue department. It seems that in many instances the village revenue officials do not allow the people to harvest the benefits of these lands. Thus, although farmers often encroach upon these lands, they are never put to rational land use (e.g. orchards, pastures, forests, woodlot or other long term use). IWDP (Hills-II) is constraints by the fact that it does not have land use-titling authority, which belongs to the district revenue office only.

Steps Taken by User Groups

The project in association with VDCs has created assets through closing degraded forestland, VCLs and constructing water-harvesting structures. The benefits from these assets have started accruing to the population of adjacent areas and transhumance communities. To maintain these assets and ensure benefits sharing, the project in association with VDCs has formed user groups such as water user groups. The constitution, number of members and the method of benefit sharing has been different for each user groups and devised by the group for its sustenance and establishing linkage with VDC. User groups have taken steps in both the forested as well as agricultural sub-watersheds to manage the CPRs. The steps taken by user groups includes developing norms for resource management, providing technical, managerial, supervisory and manpower support, conducting participatory micro-planning and monitoring exercises, besides seeking support of social groups/IWDP authorities/NGOs for the sustainable management of natural resources.

In the forested sub-watershed of Ramnagar, 81.48 per cent, 70.37 per cent of the VDC members respectively reported that they take the support of social groups/IWDP functionaries/NGOs and provide technical support in the management of CPRs. About 62.96 per cent of the VDC members provide management and manpower support for CPRs management. Similarly, in the agricultural sub-watershed of Ramnagar, 94.73 per cent of the VDC members sought the help of social groups/IWDP functionaries/ NGOs in management of CPRs. About 75 per cent of the VDC members have participated in developing norms for CPRs management in forested sub-watershed of Akhnoor, whereas in agricultural sub-watershed, only 54.54 per cent of the VDC members participated in such activities. One thing is quite clear that social groups/IWDP functionaries/NGOs have played a greater role in helping the VDCs in managing the CPRs across the selected sub-watersheds in both Ramnagar and Akhnoor.

Steps Taken for Watershed Management

The user groups have meager financial and physical resources at their disposal to perform their task efficiently. User groups have been involved to maintain and share usufruct (grass and leaf fodder) from the protected closures of forest and VCLs. The project with the help of user groups has repaired traditional drinking water sources (bowalies), irrigation water distribution channels (khuls), ponds, water-harvesting structures etc. The bowalies and ponds are mainly used for drinking and cattle population, whereas khuls are used for irrigation. The project has facilitated the transhumance nomadic and semi-nomadic pastoral communities by providing them drinking water, fodder and pasture development works, and health cover for livestock in areas where they settle during grazing periods in winter and summer. Recently, floating transhumance user groups have been formed and linked with local VDCs in selected sub-watersheds for resolution of conflicts and development of fodder banks.

User groups were active participants in the whole exercise and did play a useful role in providing implementation support in resource conservation and alternative livelihoods. The erring members were taken to tasks through social pressure and there were provisioning of assurance of rights to the resource users to elicit their responsible behaviour, for which traditional knowledge and institutions have been widely used. In some cases, local NGOs' support has also been elicited to supplement watershed development efforts at village level. None of the user groups has shared the investment needs of any kind except voluntary free labour and all the activities were carried out by the project and/or provided in VDP. The self-initiated watershed development efforts were virtually non-existing.

A majority of the VDC members (above 85 per cent) in forested sub-watershed of Ramnagar have participated in watershed management by providing implementation support and making social pressure on erring members. Whereas in the agricultural sub-watershed, more than 60 per cent of the VDC members have participated in watershed management by providing implementation support, implementing micro-enterprises through creation of SHGs, improving conservation of resources and using traditional knowledge and institutions. Similarly, more than 60 per cent of the VDC members in forested sub-watershed of Akhnoor participated in watershed management by providing implementation support, conservation activities, using traditional knowledge and support, and seeking NGOs support in their efforts.

System of User Rights

Two crucial conditions for meaningful participation in natural resource management are: sufficient incentive for the people to give their time and labour for protection in lieu of the intermediate and final products; and property rights over resources should be clearly defined in writing, well publicized and fully understood by all. Unfortunately, the usufruct sharing arrangements have been reportedly vague and not sufficiently attractive, or even relevant, for the nature of dependence on resources particularly of poor women, disadvantaged and landless. The primary focus has been on income sharing from usufructs. In all the sampled VDCs, the right of direct use of CPRs were recognized and given to the members. In some of the VDCs, right of indirect economic gain was also available. Similarly, the right of exclusion of non-members from sharing of usufruct has also been given. All the members of sampled VDCs reported that the direct use of the CPR products was given to all the stakeholders across the sampled watersheds. In the forested sub-watershed of Ramnagar, 66.66 per cent of the VDC members reported that right of indirect use of CPR product has been given to the stakeholders so that they can sell the usufructs in the market. None of the VDC has the control over the right of transfer of usufruct; however, the right of exclusion has been reported by 70 per cent of VDC members in forested sub-watersheds and by 42 and 45 per cent respectively in agricultural sub-watersheds of Ramnagar and Akhnoor.

Forest Regeneration and Protection Activities

The members of user groups participate in decision-making at village level and seem to have a considerable degree of autonomy. The user groups have framed the rules governing VCLs in general body meeting in the presence of local villagers and the project officials. User groups have the legal authority to manage the VCLs. User groups have been nested in a set of larger organizations like village *panchayats* and VDCs. The access to VCLs and forests for animal grazing has been strictly limited to local villagers. There has been provision of substantial fines for any attempt by villagers to appropriate a larger share of grazing rights by sending more animals in summer than in winter. However, these rules have not been adhered to in actual practice. Overgrazing has not been prevented. A user group has annual meetings to discuss general rules and policies and elect officials, impose fines for misuse of the VCLs, and organize the annual maintenance work. Many of the rules they use keep their monitoring and other transactions costs relatively low and reduce the potential for conflict.

User groups have been engaged actively in forest protection and regeneration activities, which have inculcated a sense of ownership among the users and created awareness on collective issues. All the members of the VDCs selected across the sampled sub-watersheds were reportedly engaged in protection of forestland. About 77 per cent of the VDC members in forested sub-watershed of Ramnagar reported that project has inculcated a sense of ownership of natural resources among them. In forested as well as agricultural sub-watersheds of Akhnoor, about 80 per cent reported that the project has created a sense of ownership among them. More than 70 per cent of the VDC members in sub-watershed of Ramnagar reported that project has created a sense of collective responsibilities among them to regenerate and protect the forest resources. However, they have been unable to evolve the system of rotational watch and ward for monitoring of resource conservation and protection activities. The project has opted for a system of local paid watch and ward that are employed with IWDP on contractual basis. The watch and ward were authorized to levy fines on those who fail to adhere to the rules; however, not a single instance of fine has been reported even for repeated failure to observe the rules.

User groups along with VDCs have created detailed authority rules specifying in various ways how much of each valued product a user could harvest from the forest and under what conditions. The rules used in selected sub-watersheds were tailored to the specific environment, to the particular economic roles that various forest products played in the local economy, and to the need to minimize the costs of monitoring labour inputs, resource unit and outputs, and compliance with the rules. User groups have also devised their own monitoring and sanctioning system. The forests usually were closed, except for specified periods; anyone caught in the enclosed forests at other times obviously was not following the rules. The forest resources have been subjected to unrestrained use and competition among users for the better spots, which had increased production cost, as well as the level of uncertainty regarding the fuel-wood and fodder potential.

All the villagers were included in the list of beneficiaries. The members were permitted to collect dried fuel, fodder and other forest resources from the protected forests on weekly basis. During winter, entry into the forests was totally banned to prevent the misuse of the forest resources and to regenerate and improve the resource potential. It is significant to note that no attempt has been made to specify the rules such as number of collector per household, distribution of cut produce and equal sharing of usufruct among the members. The individual benefit maximizing tendencies dominate in the absence of such rules. The system of rotational watch and ward seems to be unsustainable and may not persist after the withdrawal of the IWDP. The provision of fines and penalties were made to enforce the rules strictly. The defaulters were to loose their membership on user groups in case of repeated misuse of rights. The house repairs and construction requirements of the villagers are met by the use of timber, for which the villagers have to seek the prior permission of the executives of the user groups and VDCs. In such cases, the villagers have to pay a nominal fee for the timber use and strictly adhered to the rules in operation, failing which penalties and even suspension of the membership of user groups are in force. The mature trees fit for harvesting are marked by the executives and allowed to harvest by the eligible members with proper sanction of the executives of the user groups. No harvesting of trees is authorized except the repairs and construction needs.

No doubt, numerous rules have been framed for regulation of resource use, however, in actual practice the rules compliance has not been cent per cent and cases of rule infringement have also been reported. At the same time, the monitoring of rule compliance was also very difficult under the existing rules due to system of paid watch and ward. Although the existing resource system is not a private property system, rights to use forest resources and duties to respect these rights are well defined. User groups and VDCs have used the services of project functionaries to legitimize their role in helping to devise a workable set of rules. User groups and VDCs have also forged suitable linkages with line departments specifically for activities like maintenance of nurseries, tree plantation etc. For instance, more than 60 per cent of the VDC members in agricultural as well as forested sub-watersheds of Ramnagar reported that they have forged suitable linkages with line department for maintenance of nurseries and tree plantation. In agricultural sub-watershed of Akhnoor, only 45 per cent of the VDC members had done so. On the whole, the existing user groups provide an example of a self-governed common property arrangement in which the rules have been devised and modified by the participants themselves, but with external support. However, due to vested interests and inadequate capacity building of the members, the rules have not been adhered strictly and monitoring of the rules compliance has been reportedly very weak.

COST SHARING ARRANGEMENTS

Recently, election of *panchayats* has been held, which would also pave the way in motivating the beneficiaries in cost sharing. The local NGOs have also played their part in creating awareness among the stakeholders to share the cost of the project activities. Fort example in the agricultural sector, 50 per cent and 60 per cent cost of inputs respectively on rainfed crop demonstration is to be borne by the beneficiaries in project area and non-project area. As such, the difference between project area and non-project area is 10 per cent. In case of on-farm fodder production, the difference in cost sharing is much wider, as the inputs for on-farm fodder production is provided free of cost by the government department in non-project area. In project area, 50 per cent of the input costs are to be borne by the beneficiaries. It is significant to note that full labour cost in rainfed crop demonstration and on-farm fodder demonstration is to be met by the beneficiaries in project as well as non-project area (see table 1).

Table 1
Status of Cost-Sharing by Beneficiary (%)

	Cost-Sharing by Beneficiary			*Difference*
Sub-Component	*Project Document*	*Project Area*	*Non-project Area*	*Project Area - Non-project Area*
1. Agriculture				
(i). Rainfed Crop Demonstration				
a. Cost of Inputs	50	50	60	–10
b. Labour Contribution	100	100	100	0
(ii).On-Farm Fodder Production				
a. Cost of Inputs	25	50	Free	50
b. Labour Contribution	100	100	100	0
2. Horticulture*				
a. Plantation	60	60	25	35
b. Nurseries	NA	60	Free	60
3. Village Common Land*				
a. Silvipasture Development	10	25	Free	25
b. Silvipasture Maintenance	100	70	Free	70
c. Afforestation	NA	10	Free	10
d. Drainager Line	Locational **	7	Free	7
4. Water Resource*				
a. Water Harvesting Structure	25 ***	20 **	Free	20
b. Natural Water Points	25 ***	25 **	Free	25
c. Irrigation Channels	25 ***	25 **	Free	25
5. Livestock Component				
a. Artificial Insemination	8	10	Free	10
b. Natural Breeding	33	28	Free	28
c. Veterinary Services	NA	10	Free	10

Source: Status Report for Supervision Mission of World Bank, IWDP (Hills-II), J&K, 2001 *Benefit and Cost-Sharing Arrangements with Beneficiaries*, IWDP (Hils-II), J&K, 2000.
Notes: * Ratios for Labour Contribution, ** Full Maintenance by Beneficiary, and *** 50% Maintenance by Beneficiary

Over the period, the idea of cost sharing has gained momentum and beneficiaries' contribution has shown an increasing trend as revealed in table 2. For instance, in 1999-2000, 14.19 per cent of the project cost on rainfed crop demonstration was shared by the beneficiaries, which increased to 36.26 per cent in 2000-2001 and declined marginally thereafter to 34.24 per cent in 2001-2002. On-farm production, the beneficiaries' share was as high as 36.96 in 2000-2002, and declined to a low of 0.14 per cent.

This can be attributed to the fact that by 2001-2002, the project has curtailed its intervention drastically in this area, as most of the farmers have been exposed to improved methods of on-farm fodder production, and now the beneficiaries are engaged in improved on-farm fodder production activities at their own. In 2000-2001, the scheme of provisioning of vegetable kits has been implemented in a big way, for which beneficiaries contributed to the tune of 34.14 per cent of the project cost.

Table 2
Proportion of Cost Sharing by Beneficiaries on Various Sub-Component

	Cost-Sharing by Beneficiary			*Difference*
		Years		*1999-2000*
Sub-Component	*1999-2000*	*2000-01*	*2001-02*	*2001-02*
Agriculture				
Rainfed Crop Demonstration	14.19	36.26	34.24	20.05
On-Farm Fodder Demonstration	10.7	36.96	0.14	-10.56
Vegetable Kits	NA	34.14	0.25	0.25
Vegetable Field Boundaries	12.3	3.54	56.77	44.47
Horticulture				
Rainfed Horticulture Demonstration	53.45	78.34	76.46	23.01
Village Common Land				
Afforestation	4.67	4.92	10.34	5.67
Protected Forest Land/Govt. Land				
Afforestation	0	0.44	8.52	8.52
Forest Regeneration	1.56	0.62	10	8.44
Pasture Development	6.85	10.42	14.93	
Water Resource				
Water Harvesting/ *Bowlies*/Ponds/etc.	4.86	551.8	8.34	3.48
Livestock Development				
Artificial Insemination	0.008	0.07	5.189	5.181
Veterinary Services	0.0007	0.03	6.16	6.1593

Source: Status Report for Supervision Mission of World Bank, IWDP (Hills-II), J&K, 2001

BENEFIT SHARING ARRANGEMENTS

While introducing participatory watershed management, one of the first things to be mentioned is the provision for benefit sharing from plantations in enclosures on VCLs and forestlands to user groups. Not surprisingly most villagers, particularly the ones likely to be least affected by VCLs and forestlands being closed for regeneration were attracted by the prospect of receiving money in lieu of labour services. However, many project functionaries complained that the villagers are primarily interested in money but that is what they themselves use as bait to get the villagers interested. As VCLs and forestland closures take several years to regenerate, project staff finds it difficult to sustain the villagers' interest during the long waiting period. To increase its credibility among the villagers, project has started sharing usufructs from exiting older plantations with the user groups.

The members of the user groups have user rights over the CPR resources. The villagers have a customary right to collect fuel-wood and timber from the better-stocked VCLs and forests in their vicinity. They are unlikely to buy what they can get free. In any case, the poorest forest users do not have the cash to buy such subsistence goods. The issue here is not that the villagers, particularly those dependent on forest for subsistence and livelihood needs, are only interested in money but that is the only option made available. Satisfaction of livelihood or subsistence needs, especially of the poorest groups is not taken into account and the operative paradigm is usufruct sharing from non-protected VCLs and forests. The entry to the enclosed VCLs and protected forests has been restricted for the initial period of one year. After the interval of one year, the enclosures and protected areas were opened to the members for collection of dried fuel and fodder. The entry of non-members and outsiders is strictly restricted to the enclosed and protected CPRs so that others who have not contributed to those efforts will not reap any benefits produced by users' efforts. The system of paid watch and ward by the project is in place to exclude outsiders and non-members from access and appropriation rights as well as to oversee the misuse and overuse of the resources after regeneration.

CONCLUSION

Institutional change in natural resource management has come into existence having felt the need for collective action to regenerate natural resources to meet consumption needs, or to gain access to such resources for economic empowerment to emerge from existing social oppression. The institutional change has been evolved by the project and not self-initiated. The strengthening of institutional change in the formative years involves a great deal of efforts on the part of project implementing agency and local stakeholders including villagers, political leadership and NGOs. Over the period, participatory institutions have gained confidence and build their capabilities; they diversify their activities and have become more self-reliant. The creation and maintenance of closures on VCLs and forestland, and repairs and maintenance of water distribution channels and water harvesting structures with the object of meeting their needs have been the primary activity of user groups and VDCs. This initial activity has inculcated a sense of ownership and collective responsibility among members.

In nutshell, the government takes the first move to rehabilitate the degraded CPRs because they possess the resources. The local people lack the initiative to take the first move, which does not mean that a top-down approach should be followed. The process of institutional change should commence first in those villages where participation is most likely to take place. For instance, the best chances for voluntary participation can be found among the villagers who depend highly on the CPRs and perceive the quality of the forest as good. After initial performance in participatory resource management, the successful villages can then serve as an example for other villages to extend the process. Motivated by success in the first stages, resources can be mobilized to replicate the process in villages with less favorable circumstances, hence, the process should not be bottom-up either, but it should be an interaction between the state and the people, leading to a win-win situation. Transparency of the state and legal rights for the people are important aspects for success as well. In order to improve natural resource management practices, people should be given more freedom to act on their own. The state should provide resources and assistance by formally allowing them a share in rehabilitated CPRs. This would enhance the development of the village and the mutual trust between villagers, so that mutual participation can be sustained, by getting closer to the optimal natural resource management.

REFERENCES

Blomquist, W. (1992), Dividing the Waters: Governing Groundwater in Southern California, Institute for Contemporary Studies Press: San Francisco.

Bromley, D. W. (1989), *Economic Interests and Institutions: The Conceptual Foundations of Public Policy*, Basil Blackwell : London.

Chopra Kanchan and Kadekodi, G. K. (1998), *Operationalising Sustainable Development: An Ecology-Economy Model for a District in Bihar*, Sage: New Delhi.

Dasgupta, P. (1997), Environmental and Resource Economics in the World of the Poor, Invited Lecture at Resources for the Future: Washington.

Farrington, J., C. Turton and A. J. James (eds.) (1999), *Participatory Watershed Development*, Oxford University Press: New Delhi.

Farrington, J. and Lobo, C. (1997), Scaling up Participatory Watershed Development in India: Lessons from the Indo-German Watershed Development Programme, *Natural Resource Perspectives*, Overseas Development Institute: London.

Farrington, J. and Boyd, C. (1997), Scaling up the Participatory Management of Common Pool Resources, *Development Policy Review*, 4(15).

Integrated Watershed Development Project (IWDP), Hills-II (2001), *Status Report for Supervision Mission of World Bank*, IWDP (Hills-II), Jammu and Kashmir.

Integrated Watershed Development Project (IWDP), Hills-II (2000), *Benefit and Cost-Sharing Arrangements with Beneficiaries*, IWDP (Hils-II), Jammu and Kashmir.

Jodha, N. S. (1986), Common Property Resources and the Rural Poor in Dry Regions of India. *Economic and Political Weekly*, 54.

Kolavalli, S. (1998), Scaling up Participatory Approaches, Paper prepared for National Workshop for Watershed Approaches to Wastelands Development: New Delhi, 27–29 April.

Ostrom, E. (1990), *Governing the Commons: The Evolution of Institutions for Collective Action*, Cambridge University Press: Cambridge.

Quiggin, J. (1993), Common Property, Equality and Development, *World Development*, 4(21).

Sharma Prem N. (1995), An Analysis of and Lessons from Case Studies of People's Participation in Watershed Management in Asia, in *Watershed Management in Tropics and Upper Himalayas*, www. fao.org/DOCREP.

Singh, Katar (1995), The Watershed Management Approach to Sustainability of Renewable Common Pool Natural Resources: Lessons of India's Experience, *Research Paper* 14, Institute of Rural Management: Anand.

Singh Katar (1997), Property Rights and Tenures in Natural Resources, in John M. Kerr *et al.* (eds.) *Natural Resource Economics: Theory and Application in India*, Oxford and IBH Publishing Co: New Delhi.

Tang, S. Y. (1992), *Institutions and Collective Action: Self-Governance in Irrigation*, ICS Press: San Francisco, U S A.

Uphoff, N. (1993), Grassroots Organisations and NGOs in Rural Development: Opportunities with Diminishing States and Expanding Markets, *World Development*, 4(21).

Wade, R. (1988), *Village Republics: Economic Conditions for Collective Action in South India*, Cambridge University Press: Cambridge.

World Bank (1999), *Project Appraisal Document: Integrated Watershed Development (Hills-II) Project*, Rural Development Sector Unit, South Asia Region: New Delhi.

World Commission on Environment and Development (WCED) (1987), *Our Common Future*, Oxford University Press: Oxford.

29

COMMON PROPERTY RESOURCES AND SUSTAINABLE DEVELOPMENT: A PERSPECTIVE

M. Sabesh Manikandan, Dr. K. Raja Mohan Rao and S. Nehru

INTRODUCTION

The socio-economic development of man depends on the environment and this development influences the environment. Thus environment and development are interdependent. Now-a-days 'development' has become a comprehensive word carrying such wide as well as specific meanings that both the common man and the specialist can use it in their own contexts. In fact, economic development is closely linked to the development of resources. These resources can be classified into three categories, viz., public goods, semi-public goods and private goods.

In poor countries common property resources make a valuable contribution to the sustainable livelihoods of rural populations. This includes the collection of fuelwood, fodder, crop wastes, cow dung, organic manure, small timber and other products that are derived from the bark, seeds, flowers and fruits of trees, as well as water for drinking, cooking and irrigation and local fisheries. The existence of imperfect factor markets results in an intimate link between the rural economy and its natural resources base. Inadequate rural employment opportunities, especially in the slack season, imply that the local commons can make substantial contributions to household incomes. Another important function of local common property resources is that they act as insurance against uncertainty in the absence of complete contingent markets. Access to such resources serves to prevent risks associated with natural disasters and crop failure. Furthermore, for landless populations, access to local common property resources may be the only available non-human asset.

The term 'Common Property Resources' (CPRs) is broadly defined as natural resources in which a group of people has common user rights (not necessarily ownership rights). These include natural forests; village lands, grazing lands, streams, rivers, groundwater and oceans. These may also include man-made resources like irrigation tanks, community wells and village roads.

CURRENT STATUS OF CPRS IN INDIA

In India, out of the total land area of 330 mm hectares, only about 140 to 147 million hectares are cultivated. The remaining 190mm hectares, consisting of forests, woodlands, grasslands, deserts, marshes, rivers, lakes, shorelines and other forms of common properties support many occupations like forestry and livestock rearing and provide daily requirements like food, fuel, fodder and medicines.

According to one estimate, the CPR land area of the country comes to 21.55 per cent of the total geographical area (Chopra, Kadekodi and Murty, 1990). However, its magnitude varies from 9 per cent to 28 per cent of the geographical area in different districts of the country based on a selected village level survey (Jodha, 1986). Moreover, according to a recent estimate of common property resources, about 14.81 per cent of the total geographical area are non-forest common property resources in sixteen major states of India. It is very close to the 54th round survey of NSSO (1999) which reported that about 15 per cent of the common property resources in India is non–forest CPRs (Table 1).

Table 1
NSSO Estimates of Common Property Resources in India

Indicator (all India figures)	*NSSO estimates*
Share of common property resources in total geographical area	15%
Common property resources per household (in ha.)	0.31
Common property resources per person (in ha.)	0.06
Reduction in common property resource land in last 5 years (per 1000 ha.)	19 ha (0.38% p.a.)

Source: NSSO (1999)

SUSTAINABLE DEVELOPMENT

In recent years, there has been a growing recognition that sustainable development can only be achieved on the basis of a balanced appraisal of the environmental, economic and social impacts of a country's major development projects, as well as its development policies, plans and programmes.

Sustainable development is about enhancing human well being through time. One should have the ability and opportunity to shape one's life, which increase with better health, education and material comforts, enhancing the portfolio of assets. The World Commission on Environment and Development defined sustainable development as "progress that meets the needs of the present without compromising the ability of the future generations to meet their own need".

Need for Sustainable Development

Improving human well being over time is a broader goal than increasing economic growth that focuses primarily on material comforts. This has some important implications. Since, social and environmental aspects also affects human well being directly, a strict policy of 'grow now, cleanup later' involves costs for to-day's generation, costs that often fall disproportionately on 'to-day's poor'. Durable economic growth is required for a serious attempt at poverty reduction. This means paying enough attention to social and environmental concerns so that durable growth is not affected. There are limits to sustainability among assets, perhaps even more from the perspective of people's well being. When environmental or natural assets are fairly abundant relative to human made assets, substitution of the former by the latter can be expected to lead to higher returns. Development strategy to date has often relied on drawing down environmental resources and replacing them with human made assets. Most developing countries' growth strategy continues to focus largely on the accumulation of human made assets (Physical Assets).

The way an economy grows–pace and pattern of growth–can matter for the well being of both the current generation's children and grandchildren. Developing countries do not have to follow the path of development traversed in the industrial countries.

Sustainability Perspective

The imperatives of economic efficiency, ecological health and socio–political democracy are interdependent. Sustainability is:

(a) A normative standard or social goal; and,

(b) A vision of the future with many origins.

There are at least two perspectives on sustainability, namely,

(a) **The Ecological View:** In this view, sustainability is premised on the integrity of the ecosystem because there is no human–made substitute.

(b) **The Social Science View:** In this view, sustainability also requires democratic processes and a vital economy.

CPRS AND ECONOMIC OPPORTUNITIES

Income Generating Activities

Regular activities based on CPRs include the collection and sale of firewood, leaves made into plates and cups, fruits, grass for fodder, grass for thatching, honey and fish. Grass and tree fodder may also be fed to small ruminants, which can be a significant source of income, especially for the poor.

Direct Inputs to Agriculture

Wood from forests is used in making agricultural implements and bullock carts and in fencing off fields. Forage from forests and non-forest land-based CPRs is needed to feed livestock, some of which are an integral part of the agricultural system. Drinking water for livestock is another pre-requisite for their maintenance that often comes from CPRs such as rivers, village ponds and tanks.

Direct Inputs to the Home

Water and various fruits from CPRs are consumed by people, while firewood is essential for cooking food. Wood and grass for thatching are used in house construction and maintenance and wood is also used in furniture making.

A Safety Net for People in Drought Years

Since forests are relatively resilient in the face of drought, many forest-based income-generating activities can continue when crop production has failed. In addition, some communities in forest areas fell trees in extreme drought years (but not in normal years) and sell the wood for firewood to generate income. Forests and other common lands may also be a source of emergency foods such as weeds, tubers and mammals.

CPRS AND SECURITY

Food Security

The role of CPRs in food production and filling the food gap in India has increased as s result of the management and the betterment of access to common property resources, particularly in irrigation channels. The bulk of the additional production of foodgrains, especially of marketable surplus, has come mainly from irrigated areas. There has, however, been some controversy with respect to the increase in agricultural production due to irrigation because of the complementary role or seeds, fertilizers, credit pricing and other factors. It is to say that a 'with' and 'without' methodology often used for estimating the contribution of irrigation to food production shows that yields per hectare have increased substantially due to irrigation. In addition, there are gains due to increasing area under cultivation, raising cropping intensity and improvement in cropping pattern, which were made possible due to rapid development in irrigation channels. Irrigation channels are also the catalyst for the use of other inputs. Development and better management of irrigation channels are, therefore, needed for attaining the objective of food security.

In addition to that, the CPRs are ensuring the food security of the rural poor. As mentioned above the poor of purchasing power and the food security gap are fully covered by these resources. By availing of meat, milk and milk products of the live animals for their own consumption and for earning household income from the livestock animals their food security is linked to their own productivity in which CPR utilisation pattern plays a critical role.

Livelihood Security

The three primary products of the bushland are crops, trees and grass. Those primary products are often combined with other inputs (e.g. labour, capital, livestock) to provide goods and services of value to people, including: (i) human food–meat and milk from domesticated livestock, bush meat, gathered foods and cereals; (ii) energy-trees, tree products, manure; (iii) building materials–tree products, materials for thatching, material for handicrafts; (iv) conservation of biological diversity, including the special cases in which local residents benefit from tourism and safari operations; and, (v) sequestration of atmospheric carbon, especially in the roots of deep–rooted trees and grasses.

Empowerment

The poor who had significant and substantial stakes in the CPRs, which were governed within the framework of the common property management, found that when they were given private land ownership, its productivity was declining. Jodha (1986) also found that in the post Independence era, new markets opened for some of the CPR products resulting in over or unsustainable exploitation of the resources. His studies revealed that CPRs had contributed significantly towards employment and income of the poor where per household income ranged between Rs. 530 and Rs. 830 in different areas. During the Sixth Five Year Plan period (1980–85) households earning less than Rs. 4,800 per year were considered the poorest of the poor among those living below the poverty level income, which was Rs. 6,400 per annum. Thus, the proportion of income from CPRs to total income rendered between 11 and 17 per cent in the case of the poorest household (Rs. 4,800) and between and 13 per cent incase of poor households (Rs. 6,400).

Poverty Alleviation

As already discussed, the dependence of poorer household on CPRs is still a highly contested issue. It has been often argued that poor people extract more natural resources due to greater reliance on the natural resource base and also due to their high social discount rate. On the other hand, scholars posit that compared to the non-poor, the poor may depend more on the commons in relative terms, but in absolute terms their dependence is lower (Dasgupta, 1993), particularly for resources with good market opportunities. Consistent with growing theoretical literature on common-pool resource management, there is a large quantum of empirical research in India dealing with the dependence of the poor on the CPRs

(Jodha, 1986, Beck and Ghosh, 2000). Beck and Nesmith (2001) note that common property resources contribute about 12 per cent to the household income of poor rural households. In his study of common pool resources in West Bengal, India, Beck (1998) observes that access to CPR by the poor is gradually decreasing across all study villages and agro-ecological regions.

In their study of CPRs in different states of India, various authors find that CPRs contribute 1.1 per cent to 22.1 per cent, 10 per cent, 27.3 per cent and 12 per cent of the income of poorer households in Gujrat, Karnataka, Punjab and West Bengal respectively, while this figure for non–poor/ cultivating households was 0.1 to 11.4 per cent, 6.2 per cent, 22 per cent and 0.13 to 5.62 per cent in those states (Iyengar and Shukla, 1999; Singh *et al.*, 1996; Beck and Ghosh, 2000). In a study of 29 villages in South-Eastern Zimbabwe, Cavendish (1998, 1999) arrived at even larger estimates. He observes that the proportion of income based directly on the commons is about 35 per cent. Based on a qualitative assessment of babassu products in Maranho, Brazil, Hecht *et al.* (1988) also conclude that the products offer support to the poorest of the poor, especially the women.

Figure1
Sustainable Development

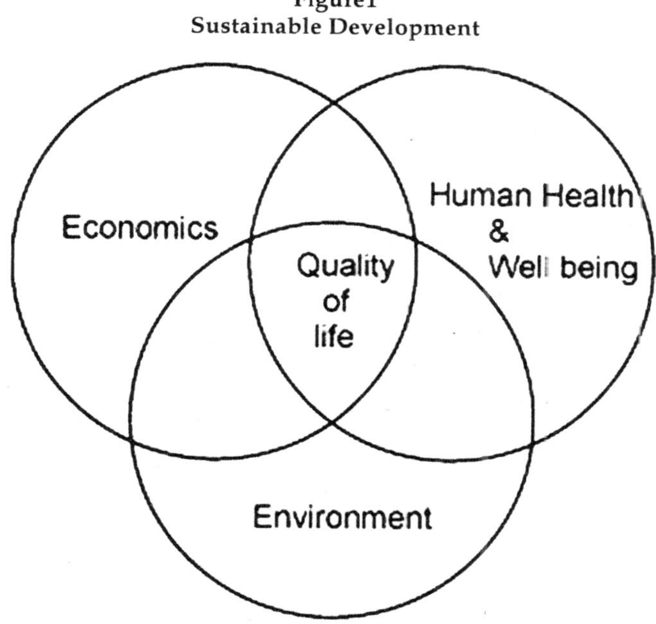

CPRS AND ENVIRONMENT

Health

The World Health Organisation (WHO) estimates that 80 per cent of all incidence of sickness in developing countries can be traced to water and sanitation related causes. The prevention of infections of such an origin is one of the most important objectives of health and sanitation programmes in India. Human excreta should be disposed of in such a manner as to avoid direct or even indirect problems. According to Central Statistical Organization, in 1993, only 3.15 per cent of the rural population had access to sanitation services. In other words, 563.6 million people living in rural areas do not have access to toilets of any type. It is a fact that disease is a challenge to a developing country like India. The poor management of the common property resources such as public toilet, drainages and the drinking water are causes for disease in the rural areas. Better management of these resources will give a pollution free and diseases free environment to the nation.

Environmental Services

Forests act as a sponge when it rains, regulating water flows, preventing flash floods and prolonging the period during which surface water is available. Where forests are on hillocks near to farmers' fields they also prevent stones and poor quality soil being washed off the hillock and deposited in farmers' fields while supplying nutrients to the fields in the form of leaf litter.

CONCLUSION

While the CPR has a unique role in providing food and food security, it shares this role with several other sectors in providing income and livelihood security. And yet, it is the income generating power of the CPR which is most important for poverty alleviation since it helps in the attainment of both CPR and food security and enables them to be self-sustaing over a long period without any resort to subsidy. Breakthroughs in productivity and income in poverty concentration areas can take place by creating CPRs. Water reservoirs of all types and sizes, watershed management recharging underground water storage etc. should be taken up with renewed vigor but without losing sight of the environment and the human aspects. Technologies and institutions appropriate to small farmers may receive special encouragement. Development of power will be needed for exploiting the groundwater potential and the operation of grain storage facilities, especially the regurgitated ones. Drainage schemes may be fed in to irrigation schemes to maintain favourable salt balance in the soil. At the same time, output from permanent CPRs like tanks, ponds, with respect to fishery and agriculture products like makhana and singhara in India should be raised by appropriate methods including regular desilting and deepening. These measures taken along with those for strengthening institutions of marketing, credit and grain storage will boost agricultural production and sustainable development in poor areas.

REFERENCES

Beck T. (1998), Excluding the Poor from their Rights: the Case of Natural Resources in West Bengal, Paper Presented at the International Association for Study on Common Property, Vancouver.

Beck T., and Nesmith, C. (2001), Building on Poor People's Capacities: the Case of Common Property Resources in India and West Africa, *World Development*, Vol. 29, No. 1: 119-133.

Beck, T. and Ghosh, M. (2000), Common Property Resources and the Poor: Findings from West Bengal, *Economic and Political Weekly*, Vol. 35, No. 3: 147-153.

Cavedish, W. (1998), *The Complexities of the Commons: Environmental Resources Demands in Rural Zimbabwe*, Centre for the Study of Aftrican Economies, Oxford University, Oxford.

Cavedish, W. (1999), *Poverty, Inequality and Environmental Resource Demands in Rural Zimbabwe*, T. H. Hukley School, Imperial College of Science: London.

Chopra, *et al.* (1990), *Participatory Development: People and Common Property Resources*, Sage Publications, New Delhi.

Dasgupta, P. (1993), *An Inquiry into Well-Being and Destitution*, Clarendon Press, Oxford.

Hetch, S.B., Andreson, A. B., and May, P. (1988), Subsidy from Nature: Shifting Cultivation, Success Ional Palm Forests and Rural Development, *Human Organisation*, Vol. 47, No. 1: 25-35.

Iyengar, S., and Shukla, N. (1999), *Common Property Land Resources in India: Some Issues in Regeneration and Management*, Gujarat Institute of Development Research, Ahmedabad, India.

Jodha, N. S. (1986), Common Property Resources and the Rural Poor in Dry Regions of India, *Economic and Political Weekly*, Vol. 21, No. 27: 169-181.

NSSO (1999), Common Property Resources in India, Results of the NSSO 54th Round, January-June, 1998, Report No. 452 (54/31/4), National Sample Survey Organisation, Government of India, New Delhi.

Singh, K., Singh, N., and Singh R. (1996), Utilisation and Development of Common Property Resources – A Field Study in Punjab, *Indian Journal of Agricultural Economics*, Vol. 51, No. 1 and 2: 244–259.

30

PARTICIPATORY FOREST MANAGEMENT: A SURVEY OF REVIEW OF LITERATURE[#]

Yogesh Ranjit

INTRODUCTION

Natural resources are the free gift of nature. Its availability (in quality, quantity and density), rational utilization and sustainable management can play a significant role on environmental protection, ecological promotion, biodiversity conservation and socio-economic development of the nation. But forests are becoming increasingly scarce in the face of pressure of them. Forests are easily available renewable natural resource. But it has been under-valued and not managed on a sustainable basis in the past. As a result, these valuable forests are declining in its availability. The estimated loss of forest area in India from 1952 to 1980 for various purposes was around 4.3 mha. of which represents at annual loss of around 1,55,000 hector (ha) per annum (GOI,1987). Similarly, during the last two decades, the percentage of forest cover in Nepal has reduced from 37 per cent to 29 per cent and the forest area has decreased at annual rate of 1.7 per cent from 1978/79 to 1994 (DoF, 1999). Such degradation of forest resources has been mainly due to the unstable government policies and lack of good governance resulting in indiscriminant encroachment of forests, extensive conversion of forest areas into agricultural land and over exploitation of forest resources both by the people and the government for commercial purposes. However, in the last three decades, the scenario has begun to change. The centralized management of forest has gradually devolved into decentralized to the local community in the name of PFM.

PFM is often defined as the practice of forestry where trees are planted on community or common land to meet the specific needs of the

[#] This paper is based on my on-going Ph.D. research work entitled 'Participatory Forest Management and Environmental-Economic Impact: A Comparative Study of Four Himalayan Districts' Department of Economics, University of Jammu, Jammu, Jammu and Kashmir, India.

community like fuel wood, fodder, grass, timber, pole and other non-timber forest products (NTFPs). However, it means, in general sense, 'Forestry of the people, for the people and by the people' (Karki, 2003). In a wider sense, it means the sustainable management and protection of existing forest, regeneration and afforestation of degraded forest and community lands through an active participation of local people and other stakeholders. The objective of each forestry activities is to ensure environmental, ecological, social, and economic gains for the local community through the improved production of biomass, soil and water conservation, local institutional development, empowerment of people (especially women, poor, disadvantaged and marginal people) etc. PFM has become more popular in late 1980s and got wider scope in 1990s. But, the model of given forest management varies from country to country based on their political, socio-economic, cultural practices like Joint Forest Management (JFM) in India, community forest management (CFM) in Nepal etc. However, they have a number of common threads.

In Nepal, perhaps first time in the world, the Government passed legislation in 1978 to handover the substantial amount of national forests in the middle hills to local communities to manage through the *panchayats*, a lowest level of local government units. Following the Decentralization Act of 1982, the focus was shifted from *panchayats* to the community forest user groups (CFUGs) with more authority and responsibility progressively devolving to these groups. This approach was given legal authority with the Forest Act-1993 and Forest Regulation-1995. Under these act and regulation, CFUGs develop their own 'Operational Plan', set the harvest rates and prices of the forest products and determine how the surplus income is determined and/or spent. Thought trees and NTFPs are legally belonging to user groups, ownership of the land remaining with the State. The state also reserved the rights to take back possession of the community forest if the terms and conditions of handover are not met. The state's role is thus defined as being that of a regulatory authority where as management control rests solely with the users of the resources. By the mid of 2006, there were 14,287 CFUGs managing 1,188,447 ha. of forests (that was about 35% of total potential area for community forestry) with the 1,642,367 total benefited households (DoF, 2006) which was around the 40% of the total households (i.e. 41,74,374) of the country (CBS, 2001). Besides, there are also a large number of informal user groups in forest management practices across the nation. Based on the Nepal–Australia Forest Project experience, the number of informal user groups existing in forest management activities in the 75 districts of Nepal is 60,720 and the average number of informal user groups per districts is about 810 (Bhatta, 1989).

But, in most communities, women as primary users/collectors of resources like fuel-wood, fodder, grass, fallen leaves and NTFPs for domestic as well as commercial purpose have very little said in decision-making process regarding protection, promotion, utilization and the management of these resources. So, the Government of Nepal issued the guidelines in 2001 namely 'Community Forestry Development Guidelines-2001', which laid particular emphasis of women participation giving mandatory to include one male and one female from each household member of user group. Consequently, the participation of women in CFM is increasing considerable. Moreover, in Nepal, there are many women CFUGs. According to the latest available data, 692 women CFUGs are solely managing 20,255 ha. of forests in 63 districts (out of 75) of the nation (Bhattarai and Khanal, 2005). However, there is a long way to go to ensure their effective participation, independent decision-making and benefit sharing in CFM.

In India, JFM got the necessary fillip in 1988 when a new forest policy was announced by the Government of India with the objective of the creation of a massive people's movement for affortation with involvement of women for achievement the policy goal of having 33% land area under forests cover. The national programme of JFM was notified in the shape of a circular from the Ministry of Environment and Forest, Government of India, issued on 1st June, 1990. But, JFM is somewhat restrictive in that it covers only the protection and maintenance of degraded forests. So, to further strengthen of the programme, the Government of India issued the new guidelines on 21st February, 2000 which envisages extension of JFM to better-stocked forests also. This guidelines laid particular emphasis of women participation (i.e. at least 50 % members of the JFM general body and 33 % of the management committee), micro-plan preparation, legal backup to the JFM committees (JFMCs), recognition of self initiated forest protection committees, resolution of conflicts, contribution for regeneration of resources (i.e. no less than 25% of the share of benefits of the village committees), effective monitoring and evaluation of the programme and performance of JFM at division and state levels.

Similarly, the Government of India again revised the guidelines and issued the new guidelines of JFM on 24th December, 2002 to further strengthen the JFM programme. It laid emphasis on bringing empowerment and accountability through a transparent memorandum of understanding singed between the Forest Department and JFMCs, non-political relationship between the JFMCs and *panchayats* and capacity building for the management of NTFPs in accordance with the micro-plans by maintaining the equity in sharing of cost and benefits, and institutional reforms. All the states and UT of India have now adopted JFM. Since 1990, as on the September 2003, 84,642 JFM committees are managing 17.33 million ha. of forests in 27 states in a varying degree of sustainable management that becomes 26 % of the total forest Area covered under JFM (Gupta, 2006).

OBJECTIVES AND METHODOLOGY

Participatory forest management has been identified as the only viable forest management strategy in third world countries. Both India and Nepal (JFM in India and CFM in Nepal) are pioneers in this field. There has been a great deal about PFM in its short history of implementation. It has been widely used, as a strategy for forest development in various parts of India and Nepal and the development in this direction has been significant in recent years. It has been successful through out the nation of India and Nepal (with some exceptional micro event). The basic objective of the paper is to show a bird's eye view scenario through the review of available literature on the PFM making different groups into international and national (India and Nepal) level.

The study is designed in accordance with the given objective so that it is, basically, based on secondary data and information which are obtained from related published and unpublished books, booklets, seminar reports, journals and other relevant literatures. The study has gone through the published or unpublished available literature and analyzed with making them into the two different groups like international and national level. The national level is also divided into the two particular countries of India and Nepal in accordance with the given objective of the study.

REVIEW OF LITERATURE

International Perspective

The late 20th century saw a dramatic transformation in global forest resources, their use and management, and people's perception of their value. Estimates showed that between 1850-1950, about 15% of the earth's resources and woodland disappeared as a result of human activities (Houghton, 1990). Since 1961, tropical countries lost over 500 million hectares (mha) of forest cover (FAO, 2000) and consumption of forest products rose by 50 per cent (Gardner-Outlaw and Engelman, 1999). The estimated annual rate of tropical deforestation was 15.5 mha. during 1980-1990 (Pandey, 1995). Between 1981-1995, the extent of the world's forests decreased by some 180 mha. representing an annual loss of 12 mha. The estimated annual loss decreased to 13.7 mha. during 1990-1995. Forests are estimated to have covered 3,454 mha. or 26.6 % of the total land area of the world in 1995 (FAO, 1997).

The role of forests in environmental protection and biodiversity became the focus of active international and local policy. At the same time, forests' critical role in the livelihoods of the poor became more widely recognized. Indeed, rural poverty is concentrated in many areas of the world's most threatened forest biodiversity and over 90 per cent of the world's poorest people depend on forests for their livelihoods (World Bank, 2001). As we enter the 21st century, the debate is intensifying, especially in developing countries, over how to reconcile the seemingly incompatible goals of conserving forests, meeting market demand, and promoting broad-based sustainable development that reduces rural poverty.

Some disenchanted observers have argued that forestry can contribute little to poverty reduction. Indeed, prevailing markets and government policies often hurt the poor. This situation has historical roots. During and following colonization, many countries promoted government-led industrialization, utilizing forests to fuel and finance public investments for national economic growth, with little concern for private market development or local economic benefits. Large-scale industrial concessions and plantations were the main forestry models, and the whole approach was based on governments claiming ownership of forests, extensively depriving poor communities or traditionally held lands. This industrialization approach not only did not benefit the poor but also achieved neither conservation nor economic development goals. And as is becoming widely recognized, where this approach remains in use, it usually gives rise to illegal logging and widespread corruption.

The 1978 *"Jakarta Declaration"* of the *'Eighth World Forestry Congress'* projected serious forest product supply gaps and warned that these could be averted only by ensuring that the economic benefits from forest utilization reach the communities owning, living, or working in the forests (Chiong-Javier, 2001). This inspired some international donors, NGOs and governments to promote small-scale commercial forestry but this enthusiasm evaporated in the face of market and policy constraints. More fundamentally, progress was constrained by the widespread perception that poor people could not manage the long-term rotations of forests; could not afford large-scale equipment; and could not supply the large volumes of wood required by pulp mills and giant sawmills. It was also widely assumed that forest enterprises needed to be vertically planned and integrated, even undertaking their own research. A lack of political will to address commercial constraints and growing concern with poverty meant that by the 1980's most external support focused on subsistence-oriented forestry (Arnold, 2001). These initiatives often helped to improve livelihood security, but they had only marginal impacts on reducing poverty. By the 1980's and 1990's, forestry projects began to include components intended to increase local cash income. But these were generally developed by public or non-profit agencies rather than the private business sector, and commonly ignored business and market realities, thus failing to achieve major income benefits (Angelsen and Wunder, 2003).

With growing concern for environment in the 1980's, overseas development assistance and development banks shifted focus dramatically from forest production to environment objectives and the establishment of protected areas became the global forestry priority. For these many reasons, low-income forest producers presently supply a small share of industrial forest products in most developing countries. In Mexico, for example, communities own 80 per cent of forests, but less than five per cent of processing capacity (Molnar and White, 2001). While millions globally are engaged in markets, they usually have low levels of output, profit and productivity. Local producers are at the bottom of a supply chain in which they lack bargaining power and technology. Local commercial forest producers are ignored by development investors, policy makers and programme planners, if not actively discouraged, to protect politically allied forest industries or forest department revenues, or for fear of negative environmental impacts.

Based on cultural practices, in Thailand, more than 10,000 patches of forests were under community managed. Local people use the forests products according to their needs but not sell in the market. The income collected from the forest products goes to the village development fund not to the user fund. Thai women are very strong in decision making so that no decisions can be made without consulting women and they keep the money and methods of the work (ICIMOD, 1995).

The South Asian countries have a tradition of managing the forests through indigenous system. Each country possesses a diverse range of cultural communities. So, the tradition of forest management practices evolved and changed over time period in response to biophysical conditions and socio-economic factors. Forests in Pakistan were thought to be free for all and there was no restriction to collect the forest products. Basically, the government centralized timber harvesting work and management. Most of the forest policy in Pakistan realized that people were the prime threats to the forests and have attempted to exclude the groups in decision-making process in those days. But about one decade back, the government of Pakistan realized the need for involving local people in forest management activities. So social / community forestry projects funded by the donor countries with people participation has been introduced by the FD through the national law. But, it covered only a small part of Pakistan's natural forests. However, there has been no major breakthrough in forest management activities (ICIMOD, 1995).

Bangladesh has 14% of forests of the total area of the nation that declined by 2% annually over the 20-year period between 1960 to 1980 (ICIMOD, 1999). So, Bangladesh's National Forest Policy of 1994 emphasized on afforestation and protection of forest with the local people participation in order to increase the forest cover from 14% to 20% of the total land area by the year 2015. Moreover, landless people have been involved in reforestation work under benefit-sharing arrangements in many programmes run by the FD. Under this programme, tenure rights was registered on the names of both husband and wife with the intension that both of them work together (Proffenburger, 1996). Community forestry in Bhutan was very much similar to CF in Nepal due to transfer of protection and management of community-used forestry to user groups with the user committee and their management plan. The natural forestry, degraded and plantation areas were handed over to user groups. Although the forest products were mostly used for substitute, 100% of the income earned through CF goes first, to the community to be used for forest management and then for community development. The forest sector makes a direct contribution of about 11% to the GDP of the country and generates about 3% of the government revenue (ICIMOD, 1995).

Forestry plays an important role in the livelihoods of hundreds of millions of rural people-principally as a subsistence safety net, but also as a source of cash income, a capital asset, and a source of employment (Sunderlin, Angelsen and Wunder, 2003). The main contribution of forest resources to rural livelihood is through providing subsistence products and services, and a *de-facto* "***safety net***." Millions of Swidden cultivators utilize forests as fallow for food crop production. Both they and farmers who practice permanent cultivation use forest foods extensively to help meet dietary shortfalls during particular seasons of the year, and during emergency periods such as floods, famines, droughts and wars (Falconer and Arnold, 1989; Scoones, Melnyk and Pretty, 1992). In dry ecosystem, open woodlands are especially critical sources of fodder for livestock herds, which provide primary subsistence and income (Kerkhof, 2000).

In West Africa, 25 per cent of people's protein requirements are met by bush meat; for indigenous groups elsewhere, it is the principal source (Bennett, 2000). Forests help farming communities to meet their needs for construction materials, household goods, fuel, animal fodder, crop nutrients and medicines. The poorest depend especially heavily on community forests. All local people rely on the environmental services of forests, particularly water quality and flow regulation (and their health impacts), habitat for crop pollinators and predators of agricultural pests, and microclimate regulation. For many people, the forest also has highly significant spiritual and religious values (Arnold and Dewees, 1995). FAO (2000) estimates that 0.1 hectare per person of forest cover is needed in low-income countries to supply these essential goods. But deforestation and population growth are reducing those critical subsistence resources. About 1.8 billion people live in 40 countries with critically low levels of forest cover.

A quarterly household economic survey was implemented with a panel of 197 households in 29 villages in Southern Zimbabwe in 1993/94 and 1996/97, with particular emphasis on environmental resource use and values. At least 100 different resource uses were identified, most notably firewood use, consumption of wild foods, livestock browse and graze, and cash income from the sale of NTFPs, such as thatching grass and carpentry products. Local producers in some areas are already actively managing their forests to produce outputs for sale (Messerschmidt, 1993; Padoch and Piñedo-Vasquez, 1996). But there are no reliable data on the aggregate income earned by rural people, especially the poor, from sale of forest products and services; as such numbers are generally not included in national statistics. Almost certainly, though, the scale is far larger than is recognized by most economists and development experts.

In Southern Ghana, 10 per cent of the population generated some cash income from forest product activities in the early 1990's. Only a minority reported that it was a major source, but more than 70 per cent stated that it was important in helping them meet particular needs or because of its timing, or in absolute terms (Townson, 1995). NTFPs play a crucial role as a source of cash income during periods of unemployment or crop failure. An assessment of the impact of the East Asian economic crisis of the late 1990's found that forest assets helped to cushion poor communities by providing supplemental income through sale of forest products, as well as a reserve of cultivable land for food production (Pagiola, 2001).

In situations where population is growing faster than per capita incomes, forest activities emerge largely to absorb people unable to obtain income, or sufficient income, from agriculture or wage employment. This situation is likely to be characterized by labor-intensive, low-return, typically household-based activities such as fuel-wood collecting and mat making. Where per capita incomes are rising, growth is more likely to be demand-driven, and low return, labour-intensive activities tend to give way to more productive and remunerative activities such as retailing, trading, and activities to meet growing and diversifying rural and urban demands. At that stage, production and selling of forest products increasingly shifts from a part-time activity of very large numbers of people to more specialized year-round operations by a smaller share of the population. In Eastern and Southern Africa, for example, woodworking for urban and rural markets grew 10 times as fast as other products, many of which are "inferior" goods. In contrast, employment in grass, cane, bamboo etc. is tied to agricultural demand and subject to high competition, and thus much less economically attractive (Byron and Arnold, 1999).

Indian Perspective

Restrictions on reckless and indiscriminate exploitation have always been the foundation of the social and cultural institutions developed by the people in various forest areas of India (Guha, 1983). There were social laws and norms that made sure that even as human beings extracted their needs from the forests, the rate of extraction did not exceed the natural growth, which avoided resource depletion. This situation changed with the first policy statement in 1894. The tribal and the forest communities then became "intruders" and "aliens over the state property". Forestlands were transformed into mere sources of revenue for the British Government

(Rangarajan, 1996) even at the expense of forest areas allocated to villagers' use (Guha and Gadgil, 1989).

After India's independence in 1947, the minimum expectation of the tribal and the social workers working among them was a basic restructuring of the forest policy and recognition of tribal rights over forests in the new forest policy of the Indian Government (Ghate, 1992). But the government of free India disappointed everybody by adopting all the basic principles laid down by the British. At the time of independence, India had almost 20 per cent of its geographical area under forest cover. The Forest Policy of 1952 envisaged bringing 1/3rd of India's geographical area under forests, with only 23 per cent of the land under the control of the Forest Department (FD). However, with the exception of protected areas and some reserved forests, which are away from human habitat, all other forests have turned into de-facto open access. According to National Remote Sensing findings, from 55.52 mha in 1972-75, the forest cover has decreased to 46.35 mha. in 1982.

This further justified the need for consolidation of authority by the state, which is very well reflected in the stringent provisions of the Forest Conservation Act-1980. However, severe protests by activists, academia, politicians, and communities at large compelled the Government to adopt a more accommodative approach, which came in the form of the Forest Policy of 1988 which envisaged bringing 1/3rd of India's geographical area under forests along with 66% and 20% of forests in hilly and plain areas respectively. The dramatic shift in the approach of government towards forest dwelling communities since 1988 has changed the expectations as well as the mutual relationship between the communities and the Forest Department. The data published by the Forest Survey of India shows that there is an increase in forest cover from 63.34 mha (19.27%) in 1997 to 77.82 mha (23.68%) in 2003 (FSI, 2005). This could be attributed to reasons like stringent laws against the conversion of forest land to non-forestry uses, deserting the populist policy of regularizing encroachment of forest lands and, especially, nation-wide voluntary efforts of communities and non-governmental organizations (NGOs) to reforest denuded lands (Baalu, 2002; Bahuguna, 2001).

Apart from these, improved techniques in evaluating forest cover and changes in the method of interpretation of data may also account for the drastic increase in the figures. Rooted in the Forest Policy of 1988, the JFM programme of the Government of India is considered an attempt to legally forge a partnership between the FD and the local community. This partnership is based on joint management objectives in which communities are expected to share in both the responsibilities as well as the benefits that would be generated. In a way, it is partially promoting common-property regimes as a means of restraining the process of forest degradation and building up a community resource base (McKean, 2000).

Rural people may earn cash income from forest resources in a variety of ways. The most widespread is through sale of NTFPs. It may account for as much as 16 per cent of total income of households in India; in Orissa, Madhya Pradesh, Himachal Pradesh and Bihar States, as much as 17 per cent of the landless depend on NTFPs to secure daily wage work, and 39 per cent are involved in NTFP collection as a subsidiary income source (Mallik, 2000). For rural women, income from non-timber forest products is particularly important. In West Bengal, a study in the early 1990's reported that three times as many women as men were involved in gathering NTFPs, that processing was exclusive to women, that twice as many women as men were involved in marketing NTFPs, and that NTFP's accounted for 20 per cent of household income (Ford Foundation, 1998).

In remote areas of Madhya Pradesh, the state forest agency, local communities and the World Bank jointly undertook a project from 1995-1999 to support JFM. The project financed technical assistance, improved planting materials, training and village eco-development projects in 2,269 forest fringe villages and served as a catalyst for funding an additional 9,603 villages by other sources. About 1.2 million households, mostly tribal people, were involved in managing over 221,000 ha. of dense forests, as well as 419,300 ha. of degraded forests. Some local people may have lost *de-facto* forest rights held earlier, but overall the project resulted in a major increase in local capital assets, as some rights to final timber harvest were transferred from the state monopoly to joint management. In addition, local people now obtain 100 per cent of the value of intermediate products and NTFPs. Payments for forest production by community members have capitalized community funds and other investments that increased agricultural productivity and opened up new income-earning opportunities which especially impressive given the associated impacts on poverty. Environmental co-benefits of the new commercial management system have been very positive. There is a significant and positive impact on the diversity and abundance of indigenous flora and fauna. Rehabilitation of severely degraded forests with indigenous vegetation was reported by villagers to reduce soil erosion, raise water tables, and increase wildlife population and biodiversity loss (World Bank, 2000). Some mahogany is now grown in plantation, but the wood quality is considered inferior (Robbins, 2000).

It is now a well-accepted fact that unless the large number of biomass-dependent communities in India is accepted as stakeholders of the forest resource, its protection is extremely difficult. This has been the basic tenet of a participatory program like JFM since 1992. However, community-initiated and NGO promoted "collective action"-based resource management had emerged sporadically throughout the country almost two decades before JFM. Studies in different parts of the country point to the existence of communities that were consciously maintaining and managing the forests within their village boundaries at their own initiatives (Guha, 1983; Sarin, 1996). In fact, in order to catalyze the processes of decentralized management, with or without government support, an informal network of NGOs too sprouted in the 1980s. This study analyses three institutional structures i.e., Self-initiated, NGO-promoted, and government sponsored JFM, evolved over time as a result of the above mentioned development in the forestry sector in India. The basic objectives of all three institutional structures have been the strengthening of ecological security and meeting the subsistence biomass needs of the local people. Yet, the three institutional structures are different, with different strengths as well as weaknesses. It would be interesting to identify the factors that may ensure the sustainability of collective action in each of these institutions.

JFM, the participatory management approach proposed by the government is not necessarily based on "collective action". It could be a forced program, which either relegates "collective action" to a paper theory or a sub-optimal level approach with no real interest in the community. Similarly, community based autonomous collective efforts suffer the likelihood of non-sustainability in the absence of legitimate sanctions set by the government. These, like the NGO-promoted efforts, can neither be replicated nor be expected to bring in change within a reasonable period of time. For that purpose, government backed efforts at larger scale are essential. However, the "participatory approach" of the government needs to lead to "collective action" by the community, which is voluntary and indigenous.

'A self-governed forest resource is one where actors, who are the major users of the forest, are involved over time in making and adapting rules within collective choice arenas regarding the inclusion or exclusion of participants, appropriation strategies, obligations of participants, monitoring and sanctioning, and conflict resolution' (Ostrom, 1997c). Apart from its effect on forest condition and optimum resource use, a shared understanding of social norms plays a crucial role in community initiated management regimes. The writings of scholars regarding commons, with historical and contemporary evidence, have shown that resource users often create institutional arrangements and management regimes that help them allocate benefit equitably, over long time periods, and with only limited efficiency losses (McKean, 1992; Ostrom, 1992a and 1992b; Agrawal, 1999. Usually communities organize themselves for collective action in the case of resources that are scarce as well as salient for the community (Gibson and Becker, 1999).

In the Indian context, NGOs have played an important role in encouraging communities to manage their own resources. They have gained credibility, especially in the forestry sector, mainly because of the unpopular regulatory role of the FD. The transition from people's ownership (free use) of the resource to the policing and regulatory approach has given a negative image to the FD amongst the people. Even after the FD adopted participatory approaches, at the initial stages the local communities found it hard to believe that the department was willing to accept them as partners in forest management. However, in many places, NGOs have played a crucial role as a facilitator in bridging the credibility gap (Varalakshmi and Kaul, 1999) as well as in providing assistance on the non-technical and social aspects of participatory forestry as distinct from the technical aspects that are more commonly handled by forestry authorities. Despite their own limitations and constraints, there is potential for the involvement of NGOs in practically every aspect of common property resource (CPR) management and in making decentralized forest management a reality. The inevitable next step after JFM is CFM, in which NGOs will have an important role to play in building community stakes in CPRs or rebuilding social capital to facilitate CPR management, or promoting bottom-up approaches to natural resource management strategies, or in facilitating the devolution process in general.

The reasons for the government's shift from a century-old centralized command and controlled management system to decentralization in the form of participatory JFM have intrigued scholars ever since its inception. Thompson (1995) best summarizes the probable reasons: fiscal crisis, exacerbated by structural adjustment, economic liberalization policies; pressure from donor agencies for greater accountability and transparency; the recognition of the failure of past approaches by state agencies; and the demonstration effect of successful pilot efforts by non-government organizations or other government agencies in other sectors.

Environmental activists and rural communities have been skeptical about the intentions of the government in sharing powers with the people, especially in the forest sector, as it is one of the revenue-generating sectors. Their apprehensions are based on past experience as well as the inherent limitations of the initial provisions of the JFM scheme. Earlier, in all government forestation programs, participation of the rural poor was largely limited to wage employment. This was in keeping with Hardin's (1968) belief that the poor destroy forests through overuse and overgrazing. The forest-poverty relationship was defined negatively, i.e., if people continue to be poor they will destroy forests. In fact, although JFM talks about the positive role that forests can play in poverty alleviation and the role that people can play in forest protection, the implementing agency, the FD itself, was apprehensive at the beginning.

Criticism against JFM has been manifold. Although JFM implied an increase in the collective ability of the communities adjacent to forests to manage, grow, and equitably share common resources, there have been few efforts to involve people in the planning process or in identifying priorities. Rarely can the communities decide which species are to be taken up for plantation. There has been little correlation between the amount of land that is brought under JFM and the amount of land required to meet the biotic requirements of people and livestock. JFM does not take into consideration the fact that the management objectives of the locals could be very different and may not coincide with those of the state (Ligon and Narain, 1999). While the JFM agreement talks about sharing long-term benefits from timber, the fate of the forests after being successfully regenerated remains hazy (Arora and Khare, 1994). Another major limitation mentioned is that the important question of "tenure" that includes clear, secure, and exclusive rights of access to the resource is kept ambiguous in the JFM scheme (Lele and Rao, 1996). Absence of legal status to Forest Protection Committees (FPCs), the government's right to dissolve FPCs, unilateral decision making, and inappropriate sharing of forest produce (Pattnaik and Dutta, 1997) are some of the other aspects criticized by scholars.

Many of these criticisms were dealt with in the two subsequent resolutions, as mentioned at the start of this paper. Recently, two states in India, Andhra Pradesh and Madhya Pradesh, have taken a step forward by decentralizing decision-making and by granting limited financial autonomy to the communities. Incidentally, both these states have been recipients of large funding from the World Bank while it is a well-known fact that 'decentralization' has been on the Bank's agenda when it comes to dispersal of funds. One is not sure whether the government is interested in expanding this shift from "joint management" to "community management" for the whole nation but one thing seems clear: it will now be difficult to go back to the days of unilateral decision-making and total control.

Similarly, the fact cannot be overlooked that more and more communities are coming forward to join JFM. By 2001 some 45,000 JFM groups were protecting approximately 12 mha. of government forests (Kumar, 2002). Only a few years ago, community management of forests was viewed as an experimental strategy. Today, it is a part of government and donor orthodoxy in forestry investment. In many parts of the world, communities living on forest fringes or within forests have regained some control over the management of these resources. However, several authors and practitioners contend that JFM falls short of its goals. The depletion and degradation of CPRs have threatened the long-term sustainability of the natural resource endowment of the country and have increased the hardship of the poor people whose basic survival invariably depends on these resources (Arnold and Stewart, 1991). The Basic causes of depletion and degradation of CPRs in western Rajasthan are the institutional changes in the form of land reform, population growth commercialization of CPRs–based activities (Jodha, 1985).

CPRs in rural Haryana are facing the serious problems of encroachment, degradation and diversion in their use due to the major factors of consolidation of landholdings, improvements in transport and communication, changes in rural economy from subsistence to market oriented, emergence of profitable markets of CPR–products especially in rural areas, widespread introduction of green revolution in crop cultivation, population growth and over exploitation of CPRs. The most discernible implications of CPR–degradation are the shortage of land, fuel wood and fodder faced by the rural area. So, according to him, the most desirable solution to this ongoing plight of people is the community-based collective management with proper participation of all local people with liberal, legal, technical and financial support of government and governmental organization (Kumar, 2000). The basic causes for deterioration of common lands in Sukhomajri and adjacent villages in Sivalik Hills of Haryanin were overgrazing, land clearance and indiscriminate tree cutting by the people (Mishra and Sarin, 1987). The real solution of the so-called *'Tragedy of Commons'* in the Sukhomajri and lower Sivalik Hills of Haryana is to establish 'effective non-governmental and non-market institutions' (Chopra *et al.*, 1990). However, illegal encroachment by rural households but the legal distribution under welfare programme as the main reasons for reduction in CPRs in the dry regions of Karnataka (Pasha, 1992).

The growth of human and livestock population, commercialization and mechanization of rural economy increased concentration of primary activities due to low diversification of rural economy, demise of craft industry etc. and that have contributed in the deterioration and further expropriation of CPRs (Singh, 1986). Forests have been exploited much beyond their sustainable rate especially for industrial and urban consumption (Guha, 1983).

The 48 per cent of the households belonging to scheduled tribes and backward communities in nine village of Ranchi district in Bihar earn their livelihood mainly by head loading–collecting fuel wood from forest and carrying it on their head to the market (Bhaduri and Surin, 1981). A large scale of deforestation in the mountainous and sub-mountainous regions where the livelihood traditionally depends has made very difficult in the collection of forest products and has increased the drudgery of women who are generally engaged in it (Kumar, 2000). Women have to spend around 3 hours of 3 days a week and to walk 5 to 10 km in fuel-wood collection for domestic use (Agarwal, 1989).

The role of government (state) as a "conservator of forests" in forest use and management in Karnataka (Uttarakhanda and Shimoga) is to regulate the use, promote and regeneration of forest and environmental care in forests treated as reserved. Because, local people hardly raised any environmental issues and gave any evidence of such awareness. (Nadkarni, 1989). The government alone can not successfully do the forest management programmes in Uttarakhand without involving people so that the strong support of local institutions and NGOs is quite necessary in this regard. The lack of institutional participation at the local level is a great constraint on forest development. Therefore, such institution will have to be helped to evolve and strengthen through appropriate programmes (Mehta, 2001).

Nepalese Perspective

Nepal has made a significant progress in development of forest resources through the community forestry programme (CFP). As compared to the past efforts in forest development, the CFP gained its momentum after it was initiated in late 70's due to its nature of operation and procedures (Joshi, 2004). The Community Forest Programme (CFP) in Nepal is successful in increasing the greenery of degraded sites, biodiversity and environmental situation forming, local level institutions for revenue management and improving the supply of forest products to farmers in the Hills of Nepal (Acharya, 2003). The community forestry intervention has contributed greatly to the development of the forest resource management institutions at the gross-root level in Nepal (Malla, 2000). Community forestry is clearly contributing to rural peoples' livelihood through the acquisition of resources from forests products and other resources (Allison *et al.*, 2004). There is a vital importance of bottom-up planning in the community forestry that involves and consults each and every stakeholder so that one can guarantee that the principles of good governance, transparency and sustainability are met (Joshi, 2004).

Forestry sector of Nepal accounts for 15 per cent of the national GDP, providing an annual employment to 17.8 per cent of the economically active population (HMG/N, 1988). Though it is not a substantial to the national GDP in comparison to the contribution from the agricultural sector. It is also estimated that about 61 per cent of total forest (5.5 mha.) in Nepal is the potential community forest (Singh, 2004). Forest products are the major source of income of CFUGs, which constitutes about 82 per cent of their total income and CF is probably contributing about NRs. 2 billion of Nepal's GDP through forest product alone. In many parts of Nepal, up to a quarter of the total household income is derived from the sale of NTFPs (Malla, 2000). Besides, community development comprises the highest proportion of CFUGs expenses (36%) that includes school support, road construction, and other community infrastructure development (Kandel and Niraula, 2004). Rural Community still consumes 69 per cent energy from fuel-wood from community forestry and this percentage has not decreased (Shrestha and Sharma, 2004).

As the process of CF in Nepal has been progressed towards maturity, new generation issues have emerged out like good governance, rural livelihood and sustainable development. Although community forestry is a nation wide programme, its most challenging tasks are to address the issue of poor, women and disadvantaged groups and its pro-poor programme is sporadic only in few CFUGs (Singh, 2004). Next challenge of CF is to enhance the productivity of CF to ensure the equitable distribution of its benefits through the transformation of natural resources into other assets that can address the livelihood priorities of CFUGs particularly those of the poorest (Allison *et al.*, 2004). The success of the CFP rests with the CFUGs formation process in which strategic procedures need to be designed and promoted to ensure their proper representation and to help them to understand their authority, responsibility and accountability. Hence, a re-orientation of the CFP is required in order to face challenges of linking community forestry with poverty reduction, livelihood promotion, income generation, good governance and sustainable forest management (Kanel and Kandel, 2004).

The forest use and management would be integrated with strategies of economic development, resources use and due recognition of the role of environment in sustainable development which is possible only through the active cooperation and participation of the local people (Nadkarni, 1989). There is a vital importance of sustainable community forest management (SCFM) in order to protect environmental degradation, conserve and promote bio-diversity, maintain eco-system, improve livelihood of forest user groups and to enhance social capital. Moreover a model community forest has to be established and developed in each district for good governance, livelihood improvement, social equity and active forest management through which CFUGs can learn and demonstrate a good example from others (Shrestha and Sharma, 2004). The transformation of the existing protection-oriented management approach to the active production-oriented management approach of community forestry which could help to fulfill the diverse need of local community on the one hand, where as, on the other, it supports maintaining biodiversity conservation for a long run even in the absence of formally established forest protected area and such a strategy ultimately supports poverty alleviation programme in Nepal (Sigdel, 2004).

The management of FUGs has been increasingly focused on community development and poverty reduction activities. But the poor people who basically depend on community forest are exploited. Hence, there is a need to help the poor in exploring better opportunities to improve their livelihoods. Besides, their access to credit, skills and marketing services has to be improved (Kandel and Subedi, 2004). Equitable distribution of forest products has not been practiced because of less involvement of poor, women, landless and disadvantaged group in decision making (Shrestha and Sharma, 2004). Consequently, one class may corner the economic benefits of the forest use, while the environmental costs and problems may be borne by another. The richest class of households gains the most from community forestry, the middle income classes gain less than the rich and poorest households gain the least and producing environmental goods from forest is mainly a female activity (Adhakari, 2003). A question may be raised regarding the allocation of CF land to the poor. Elite members in particular may not be interested to allocate CF land to the poor. So, in general, a very small portion of the CF land is allocated to the poor for income generating activities and only diverse types of CF land have been found allocated to the poor. Because, no specific rules, regulation and guidelines have been prepared mentioning the quality and quantity of CF land to be allocated for the poor (Kandel and Subedi, 2004). Besides, the additional challenge of sustainable development approach of CF is to minimize the gaps between policies and practice (Allison *et al*, 2004). Moreover, redefining of policy objectives from basic needs to poverty alleviation is also the challenge for CFM programme in Nepal.

Arnold (2001) notes that "For some time, the main thrusts of forestry development and aid strategies, while doing much to strengthen forestry's role in providing a safety net for the poor, are doing much less to help the latter cope with the opportunities and pressures that confront them from increasing liberalization, privatization and exposure to the market". Malla, (2000) has decried the lost opportunities for poverty reduction in Nepal due to continued restrictions limiting community forest user groups' market activities. Because few governments or industries monitor forest production from low-income producers (and because so much of that production is technically illegal), even vibrant local forest commerce is largely 'invisible', as are its local income and employment multiplier effects. Unless a more concerted and more ambitious effort is made, the poor will continue to lose out and the many fine efforts of so many actors in forest development will have relatively limited impact.

The poverty reduction is the basic promise of the Tenth Five Year Plan (2002-07) of the Government of Nepal. If forest resources are properly managed and utilized in favour of poor and disadvantaged people, it can help a lot to reduce poverty (Arnold, 2001).

COMPARATIVE ANALYSIS OF PFM

Similarities

Basically, in India and Nepal, forests were generally considered as the source of government revenue. The growth of human and livestock population, commercialization and mechanization of rural economy have further contributed in the process of deforestation. Forests have been exploited much beyond their sustainable rate especially for industrial and urban consumption that has made very difficult to rural women as primary forest users. Many countries in the World including India and Nepal, still pursing indigenous forest management system based on their biophysical and socio-economic condition. Worldwide area of forest cover, including in India and Nepal, decreased basically during 1950's to 1980's. The forest management and their uses shifted from government control to People's participation and also shifted from forest production to environmental and ecological objectives in the 1980's.

Forestry plays a significant role in the livelihoods of rural people as a source of fuel wood, fodder, grass, fallen leaves, cash income and employment. Rural people could earn cash income through the sale of NTFP's that may account for 20% to 25% of household income. The forest management cannot be successfully done without active cooperation and participation of the local people in which NGOs could play a crucial role as a facilitator to encourage local people and manage resource. There is no equal distribution of forest benefits due to less involvement of poor, women, landless and disadvantaged groups in decision making which is the most challenging tasks in participatory forest management system. The participation of women in forest management and decision-making is still low but gradually increasing.

Dissimilarities

In Nepal, all forest benefits belong to the forest user groups and no need to share it to the Forest Department. But in India, benefit sharing varies from state to state. Generally, all benefits from NTFP's go to the FPCs and only 25% of income from timber and poles goes to FPCs and rest 75% goes to Forest Department. The uses and management of forests have been integrated with the economic development and poverty alleviation program in Nepal that is still lag behind in Indian context.

In Nepal, the focus has been highly concentrated on involvement of poor, women, landless, and disadvantaged group in forest management and decision-making and equally distribution of forest benefits among them. Besides, livelihood, good governance, gender and equity and sustainable development also took place in priority as the second-generation issues in the 21st century. Similarly, it has also been developed the concept of model community forest to be established through which others could learn a lot. The pace of people's participation is also somehow less in India. But the use of NTFPs in rural livelihood in higher in India which could be lesson for others.

Box 1
Studies Related to PFM

Studies.	*Findings*
International	
FAO, 1997.	Between 1981-1995, the extent of the world's forests decreased by some 180 mha. representing an annual loss of 12 mha.
World Bank, 2001.	Over 90 per cent of the world's poorest people and rural poverty depend on forests for their livelihoods.
Houghton, 1990.	There was a dramatic transformation in global forest management and their uses through the people's participation since the late 20th century and the global forestry priority dramatically shifted from forest production to environmental and ecological objectives in the 1980's.
Arnold, 2001.	Most external support by the 1980's focused on subsistence-oriented forestry.
ICIMOD, 1995.	South Asian countries still have been pursuing indigenous forests management system based on their biophysical and socio-economic condition. There has been no major breakthrough in forest management activities in Pakistan. The participation of women in forest management and decision-making is low but gradually increasing.
Sunderlin, Angelsen and Wunder, 2003.	Forestry plays an important role in the livelihoods of rural people as a source of cash income, capital asset and employment.
Thompson, 1995.	The main reason for shifting top-down to bottom-up approach is the failure of past approaches by state agencies.
India	
FSI, 2005.	From 1987 to 1997, percentage of forest cover decreased from 19.49 per cent to 19.27 per cent.
Rangarajan, 1996.	Forestlands were transformed into mere sources of revenue for the British India.
Ghate, 1992.	The independent India adopted all the basic principles of forest management laid down by the British India.
Bahuguna, 2001.	The dramatic shifting of government approach towards forest dwelling communities and the mutual relationship between the communities and the Forest Department has changed since 1988. The increase in forests area at present is due to stringent laws related to forests and nation-wide voluntary efforts of communities and NGOs to reforest denuded lands.
Malik, 2000.	Rural people could earn cash income through the sale of NTFPs, which may account for as much as 16 per cent of total income of households in India.
Ford Foundation, 1998.	In West Bengal, many women were involved in gathering, processing and marketing of NTFPs, which accounted for 20 per cent of household income.
Varalakshmi and Kaul, 1999.	NGOs have played a crucial role in encouraging communities as a facilitator to manage their own resources and they have gained credibility, especially in the forestry sector.

Lele and Rao, 1996.	The legal status to Forest Protection Committees should be very much clear and secure for development of JFM in all aspects.
Kumar, 2000.	The most discernible implications of CPR – degradation are the shortage of land for agricultural, fuel wood and fodder faced by the rural area so that the most desirable solution to this ongoing plight of people is the community-based collective management with proper participation of all local people with legal, liberal, technical and financial support of government and governmental organizations.
Singh, 1986.	The growth of human and livestock population, commercialization and mechanization of rural economy have contributed in the deterioration and further expropriation of CPRs.
Guha, 1983.	Forests have been exploited much beyond their sustainable rate especially for industrial and urban consumption.
Kumar, 2000.	A large scale of deforestation in the mountainous and sub-mountainous regions has made very difficult in the collection of forest products and has increased the drudgery of women who are generally engaged in it.
Mehta, 2001.	The forest management programmes can not be successfully done without involving people and strong support of local institutions and NGOs.
Nepal	
DoF, 1999.	During the last two decades, the percentage of forest cover in Nepal has reduced from 37 per cent to 29 per cent and the forest area has decreased at annual rate of 1.7 per cent from 1978/79 to 1994.
Joshi, 2004.	Nepal has made a significant progress in community forestry programme (CFP) due to its nature of operation and procedures.
Allison *et al.*, 2004.	Community forestry is clearly contributing to rural peoples' livelihood.
Joshi, 2004.	There is a vital importance of bottom-up planning in the community forestry in other to meet good governance, transparency and sustainability forest management.
Malla, 2000.	In many parts of Nepal, up to a quarter of the total household income is derived from the sale of NTFPs.
Shrestha and Sharma, 2004.	Rural Community still consumes 69 per cent energy from fuel-wood from community forestry and this percentage has not decreased.
Nadkarni, 1989.	The integration of forest use and management with strategies of economic development is possible only through the active cooperation and participation of the local people.
Adhakari, 2003.	The richest class of households gains the most from community forestry and poorest households gain the least and producing environmental goods from forest is mainly a female activity.

Shrestha and Sharma, 2004.	Forest products have not been equally distribution due to less involvement of poor, women, and landless and disadvantaged group in decision-making.
Kandel and Subedi, 2004.	The poor people who basically depend on community forests are in need of better opportunities to improve their livelihoods by providing their access to credit, skills and marketing services.
Kandel and Subedi, 2004.	In general, a very small portion of the CF land is allocated to the poor for income generating activities and only diverse types of CF land have been found allocated to them due to the lack of specific rules, regulation and guidelines in this regard.
Sigdel, 2004.	The active production-oriented forest management approach could help to fulfill the diverse needs of local community and support maintaining biodiversity conservation that ultimately supports poverty alleviation programme in Nepal.
Arnold, 2001.	If forest resources are properly managed and utilized in favour of poor and disadvantaged people, it can help to reduce poverty.
Shrestha and Sharma, 2004.	A model community forest has to be established and developed in each district through which others could learn a lot.
Singh, 2004.	The most challenging tasks in community forestry are to address the issue of poor, women and disadvantaged groups.
Allison *et al*, 2004.	The next challenge of CF is to enhance the productivity of CF and equitable distribution of its benefits among the user group members. Besides, the additional challenges for sustainable development of CFM programme in Nepal are redefining of policy objectives from basic needs to poverty alleviation and to minimize the gaps between policies and practice.

CONCLUSION

Forests play a significant role on environmental protection, ecological promotion, biodiversity conservation and socio-economic development of the nation. But there was a gradually decrease in forests in most of the countries in the world basically in the South Asian countries from 1850 to 1950. The pace of decreasing forests was at the apex level between 1950 to 1980 and it was more severe in India and Nepal. Indiscriminate encroachment of forests for agriculture and other non-forest uses, increasing demand for timber, fuel-wood, fodder, grass, grazing land and NTFPs, increasing process of urbanization, industrialization, commercial logging, development of infrastructure and eradication of malaria are perceived as direct causes of deforestation. But these causes are driven by the population growth (both human and livestock), rural poverty, illiteracy, mass rural unemployment, extension of market, policy failures (due to unsustainable government forest policies and legislation), lack of good governance and these causal factors are influenced by socio-economic, cultural, political and some other factors in a broader context. It is no doubt that such complex problems could not be solved by the traditional forest management practices with the century-old centralized forest management policy of the government.

By the 1970's, the environmental and ecological effects were being felt with the incidence of floods, landslides, droughts, global warming, green house effects, ozone layer depletion etc. At the same time, the 1978 *"Jakarta Declaration"* projected the importance of forests for environmental protection, ecological promotion, biodiversity conservation and socio-economic development of the nation. So, in the late 1970s, under the principal of "forestry of the people, for the people and by the people", government of the most of the countries shifted forest management policy from centralized to decentralized to the local community in the name of PFM. But there are different models in different countries based on their political, socio-economic and cultural practices etc.

There was a dramatic shift in the Government approach towards forest management, people living in and around the forests, and the relationship between the forest dwelling communities and FD in the late 1980's. In the early 1990's, both governments of India and Nepal considered an attempt to legally forge a participatory forest management in which local communities are responsible for forest protection and management for environmental protection, ecological promotion, biodiversity conservation and income generation with sustainable development of forests. But in many parts of Nepal and in tribal regions of many states of India, indigenous forest protection and management groups are still functioning and formally recognized community forest management groups also differ by project funded in Nepal and by states in India. Moreover, both countries have different pattern of community involvement in forest management. So, in this ground, both countries need to think more creatively about how to incorporate indigenous forest management systems with their programmes.

The process of forest management, utilization, and sustainable development should be integrated with the strategies of socio-economic development programmes which is possible only through the active cooperation and participation of local people especially women, poor and disadvantaged communities with the technical support of the FD. Moreover, the role of NGOs and other stakeholders as advisor and facilitator should not be ignored in those programmes. But they have to clearly understand at first about the needs and wants of the local communities and interests of all NGOs themselves. However, the sharing of experiences between India and Nepal can only strengthen the future of JFM and CFM. Hopefully, this work will serve and help researchers, readers, forest managers, NGOs workers and community members etc. to get some ideas and exchange information between the two countries.

REFERENCES

Acharya, K. P. (2003), 'Twenty-four years of community forestry in Nepal', Paper presented in a project evaluation workshop on community forestry, DoF, HMG, Kathmandu, Nepal.

Adhikari, B. (2003), 'Property rights and natural resources: socio-economic heterogeneity and distributional implications of common property resource management', Working paper, SANDFE, Kathmandu, Nepal, No. 1-03.

Agrawal, A. (1999), *Greener Pastures: Politics, Markets and Community Among a Migrant Pastoral People,* Durham, NC, Duke University Press.

Agarwal, B. (1989), 'Rural woman, poverty and natural resources: substance, sustainability and struggle for change', *Economic and Political Weekly,* Vol. 24, No. 43,

Allison, G., J. Bampton, B. R. Kanel, M. L. Shrestha, N. K. Shrestha (2004), 'Community forestry and livelihoods: how can community forestry better contribute to the Millennium Development Goals', Paper presented at the Fourth National Workshop on Community Forestry, DoF, Kathmandu, Nepal.

Angelsen, A. and S. Wunder (2003), Exploring *the Forest-Poverty Link: Key Concepts, Issues and Research Implications,* Center for International Forestry Research, Bogor, Indonesia.

Arnold, J. E. M., and W. C. Stewart (1991), 'Common property resource management in India: Tropical forest paper, Oxford Forestry Institution, University of Oxford, Oxford, No. 24.

Arnold, J. E. M. (2001), Forestry*, Poverty and Aid,* CIFOR, Occasional Paper, Center for International Forestry Research, Bogor, Indonesia, No. 33.

Arnold, J. E. M and P. A. Dewees (1995), *Tree Management in Farmer Strategies: Responses to Agricultural Intensification,* Oxford University Press, New York.

Arora, H. and A. Khare (1994), *Experience with the Recent Joint Forest Management Approach,* Paper prepared for the IN workshop on India's Forest Management and Ecological Revival, New Delhi, India.

Baalu, T. T. R. (2002), *Speech at the Function for Launching of Ecology of Hope Initiative in Johannesberg,* South Africa.

Bahuguna V. K. (2001), *Production, Protection and Participation in Forest Management: An Indian Perspective in Achieving the Balance,* key-note address at the 16th Common Wealth Conference, Perth, Australia, published in the conference proceedings-ttp://www.envindia.com/FORBAHUGV.htm).

Bennett, E. (2000), *Timber Certification: Where is the Voice of the Biologist? Conservation Biology,* 14(4), 921-923.

Bhaduri, T. and V. Surin (1981), 'Community forestry and women head-loaders', Paper presented in a seminar on role of women in community forestry. FRI, Dehradun.

Bhatta, B. R. (1989), 'Community and Private Forestry in Nepal: The Institutional Issues', Paper prepared for ICIMOD-UNESCO-MAB Symposium on Mountain Environmental Management, April 11-14, 1989, ICIMOD, Kathmandu, Nepal.

Bhattarai, A. M. and D. R. Khanal, (2005), *'Communities, Forests and Law of Nepal: Present State and Challenges'*, Federation of Community Forestry Users (FECOFUN), Kathmandu, Nepal.

Byron, N. and J. E. M. Arnold (1999), 'What Futures for the People of Tropical Forests?' *World Development*, 27(5), pp. 789-805.

CBS, (2001), 'Statistical Year Book', National Planning Commission Secretariat, HMG, Kathmandu, Nepal.

Chiong-Javier, M. (2001),'Local Organizations in Upland Natural Resource Management in the Philippines: Country Overview', Paper presented at the SANREM Conference on 'Local Governance of Natural Resource Management in Southeast Asia', Manila, May, 28-30.

Chopra, K., G. K. Kadekodi, and M. N. Murty (1990), *'Participatory Development: People and Common Property Resources,* Sage Publication, New Delhi.

DoF (1999), Forest *Resources of Nepal (1987-1998), Department* of Forest Research and Survey, MoFSC, Government of Nepal, Kathmandu.

DoF (2006), FUG Database Record available in MIS, Community Forest Division, Department of Forest, Ministry of Forests and Soil Conservation, Government of Nepal, Kathmandu.

Falconer, J. and J. E. M. Arnold (1989), 'Household Food Security and Forestry: An Analysis of Socio-Economic Issues', *Community Forestry,* Note 1. Food and Agriculture.

FAO (1997), *State of the world Forests,* FAO, Rome. Italy.

FAO (1999), *State of the world Forests,* FAO, Rome. Italy.

FAO (2000), 'Development of National Level Criteria and Indicators for the Sustainable Management of Dry Forest of Asia', Background paper, *Asia and Pacific Forestry Commission,* RAP Publication, Bhopal, India, 30 Nov-3 Dec.

Ford Foundation (1998), *Forestry for Sustainable Rural Development: A Review of Ford Foundation-Supported Community Forestry Programme in Asia,* Ford Foundation, New York.

FSI (2005), State *of Forest Report-2003,* Forest Survey of India, Ministry of Environment and Forest, Dehra Dun, India.

Gardner-Outlaw, T. and R. Engelmann (1999), Forest *Futures: Population, Consumption and Wood Resources,* Population Action International, Washington, D.C.

Ghate, R. (1992), Forest *Policy and Tribal Development,* New Delhi, India, Concept Publishing Company.

Gibson, C. and D. Becker (1999), 'The Lack of Institutional Demand', in Gibson, Ostrom and McKean, eds, *People and Forests: Communities, Institutions and Governance,* Cambridge, MIT Press, 135-162.

GoI (1987), India's Forests Survey and Utilization Division, MoEF, New Delhi, India.

Guha, R. (1983), 'Forestry in British and Post-British India: a Historical Analysis', *Economic and Political Weekly,* Vol. 18, No. 44 and 45, pp. 1882-1892 and 1940-1947.

Guha, R. and M. Gadgil (1989), 'State Forestry and Social Conflicts in British India', *Past and Present,* 123 (May): 141-177.

Gupta, H. K. (2006), *'Joint Forest Management: Policies, Participation, and Practices in India'.* International Book Distributors, Dehradun, Uttaranchal, India.

Hardin, G. (1968), Tragedy *of the Commons,* Science, Vol. 162, No. 3859, pp. 1243-1248.

HMG/N (1978), 'Panchayat Forestry Rules and Panchayat Protected Forest Rules', *Nepal Rajpatra (Gazette),* His Majesty's Government, Kathmandu, Nepal.

HMG/N (1982), 'Decentralization Act-1982', His Majesty's Government, Kathmandu, Nepal.

HMG/N (1988), *'Master Plan for the Forestry Sector',* His Majesty's Government, Ministry of Forest and Soil Conservation, , Kathmandu, Nepal.

HMG/N (1993), 'Forest Act-1993', His Majesty's Government, MoFSC, Kathmandu, Nepal.

HMG/N (1995), 'Forest Regulation-1995' Government of Nepal, MoFSC, Kathmandu, Nepal.

Houghton, R. A. (1990), *'The Global Effects to Tropical Deforestation,* Environmental Science and Technology', 24(4): 414-424.

ICIMOD (1995), *Community Forestry: The Language of Life;* Report of the First Community Forestry Users' Group workshop; organized by ICIMOD and FAO's Forest, Trees and People Programme, Kathmandu, Nepal, 22-27 May, 1995.

ICIMOD (1999), *Participatory Forest Management: Implications for Policy and Human Resources' Development in the Hindu Kush-Himalayas;* Workshop Proceeding, eds.A. Bhatia and S. Karki, Vol. 1, Kathmandu, Nepal.

Jodha, N. S. (1985), 'Population growth and the decline of common property resources in Rajasthan, India', *Population and Development Review,* Vol. 11, No. 2, pp. 247-264.

Joshi, D. B. (2004), 'Programme planning in community forestry', *Community Forestry Bulletin,* DoF, HMG, Kathmandu, Nepal.Vol. 11, pp. 20-33.

Kandel, B. R. and R. Subedi (2004), 'Pro-poor community forestry some initiatives from the field', Paper presented at the 4th National Workshop on community Forestry, DoF, HMG, Kathmandu, Nepal.

Kandel, K. R. and D. R. Niraula (2004), 'Can Rural Livelihood be Improved in Nepal through Community Forest? Banko Janakari, DoF, HMG, Kathmandu, Nepal, Vol. 14, No. 1.

Kanel, K. R and B. R. Kendel (2004), 'Revising community forestry in Nepal: achievements and challenges', *Community Forestry Bulletin*, DoF, HMG, Kathmandu, Nepal, Vol. 11, pp. 5-19.

Karki, M. (2003), Joint *and Comm8unity Forest in South Asia: Lesson Learned, Road Blocks and Future Prospect,*. Paper presented at the Joint Forest Management in Pakistan. Peshawar, July 30-31, 2003.

Kerkhof , P. (2000), *Local Forest Management in the Sahel*, SOS, Sahel, London.

Kumar, S. (2000), *'Common Prosperity, Community Interests and Environmental Concerns'*, Manak Publication, New Delhi, India.

Kumar, (2004), ' Does "Participation" in Common resource Management Help the Poor', a Social Cost-benefit Analysis of Joint Forest Management in Jharkhand, India, World Development, 30: 763-782.

Lele, S. and R. J. Rao (1996), 'Whose Co-operatives and Whose Produce? The Case of LAMPS in Karnataka', Paper presented at the National Seminar on 'Rediscovering Co-operation', November, 19-21, Institute of Rural Management, Anand.

Ligon, E. and U. Narain (1999), 'Government Management of Village Commons: Comparing Two Forest Policies', *Journal of Environmental Economics and Management*, 37(3), 272-289.

Malla, Y. B. (2000), *Impact of Community Forestry Policy on Rural Livelihoods and Food Security in Nepal, Unasylva*, 51(202): 37-45.

Mallik, R. M. (2000), *Sustainable Management of Non-timber Forest Products in Orissa: Some Issues and Options, Indian Journal of Agricultural Economics*, 55(3). July-Sept, 383-397.

McKean, M. A. (1992), 'Management of Traditional Common Lands (Iriaichi) in Japan', in D. Bromley, ed., *Making the Commons Work: Theory, Practice and Policy*, San Francisco, Institute for Contemporary Studies, 63-93.

McKean, M. A. (2000), 'Common Property: What Is It, What Is It Good for, and What Makes it Work?', in C. Gibson, M. McKean, and E. Ostrom, eds., *People and Forests: Communities, Institutions, and Governance*, Cambridge, MA: MIT Press.

Mehta, K. S. (2001), 'Participatory in forest management: a case study of Uttarkhand', in G.P. Mishra and B.K. Bajpai, (eds.) *Community Participation in Natural Resource Management*, Rewat publication, Jaipur, India, pp. 315-328.

Messerschmidt, D. A. (1993), *Common Forest Resource Management: Annotated Bibliography of Asia, Africa, and Latin America*, Food and Agriculture Organization, Rome, Italy.

Mishra P. R. and M. Sarin, (1987), 'Sukhomajri Nada a new model of eco-development', *Business India*, Issue in 253, pp. 78-80.

Molnar, A. and T. A. White (2001), 'Forestry and Land Management', in M.M. Guigale, O. Lafourcade and V. H. Nguyen, eds. *Mexico: A Comprehensive Development Agenda for the New Era,* The World Bank, Washington, D.C., pp. 669-681.

Nadkarni, M. V. (1989), *'The Political Economy of Forest Use and Management',* Sage Publication, New Delhi, India.

Ostrom, E. (1992a), *Crafting Institutions for Self-Governing Irrigation Systems,* San Francisco, CA, ICS Press.

Ostrom, E. (1992b), 'Policy Analysis of Collective Action and Self-Governance', in W.N. Dunn and R.M. Kelly, eds. *Advances in Policy Studies Since 1950, Policy Studies Review,* Annual 10 , 81-119.

Ostrom, E. (1997c), 'Self-Governance and Forest Resources', Conference on 'Local Institutions for Forest Management: How Can Research Make a Difference,' CIFOR, Bogor, Indonesia, November 19-21. http://www.cgiar.org/cifor.

Padoch, C. and M. Pinedo-Vasquez (1996), ' Smallholder Forest Management: Looking Beyond Non-Timber Forest Products', in M. Perez, and J. E. M. Arnold, eds. *Current Issues in Non-Timber Forest Products Research,* Center for International Forestry Research, Bogor, Indonesia, pp. 103-118.

Pagiola, S. (2001), *Deforestation and Land Use Changes Induced by the East Asian Economic Crisis,* EASES (East Asia Environment and Social Development Unit) Discussion Paper, March 2001, World Bank, Washington, D.C.

Pandey, D. (1995), *Forest Resource assessment 1990:Tropical Forest Plantation Resources,* FAO Forestry Paper No.128, FAO, Rome.

Pasha, S. A. (1992), 'Common Property resources and poor: a micro level analysis', *Economic and Political,* weekly, Vo. 27, No. 46, pp. 2499-2503.

Pattnaik, B. K. and S. Dutta (1997), 'JFM in Southwest Bengal: a Study in Participatory Development', Economic *and Political Weekly,* December 13, 1997.

Proffenberger, M. (1996). (ed.), Communities and Forest Management: A report of the IUCN Working Group on Community Investment in Forest Management with Recommendation to the Intergovernmental Panel on Forests. Washington: IUCN.

Rangarajan, M. (1996), Fencing *the Forest: Conservation and Ecological Change in India's Central Province, 1860-1914,* New Delhi, Oxford University Press.

Robbins, C. S. (2000), *Mahogany Matters: The U.S. Market for Bio-Tropical Mahogany and Its Implications for the Conservation of the Species,* TRAFFIC, USA, Washington, D.C.

Sarin, M.(1996) (reprinted, 1998), 'From Conflict to Collaboration: Institutional Issues in Community Management', in M. Poffenberger and B. McKean, eds., *Village Voices, Forest Choices,* Delhi, Oxford University Press, 165-203.

Scoones, I., M. Melnyk and J. N. Pretty (1992),' The Hidden Harvest: Wild Foods and Agricultural Systems', *A Literature Review and Annotated Bibliography*, International Institute for Environment and Development, London.

Shrestha, R. B. and A. R. Sharma (2004), 'Sustainable management of community forests: contribution towards Millennium Development Goals', Paper presented to the 4th National Workshop on community forestry, 4-6 August, DoF, HMG, Kathmandu, Nepal.

Sigdel, E. R. (2004)', 'Community forestry for bio-diversity conservation and poverty alleviation', *Community Forestry Bulletin*, DoF, HMG, Kathmandu, Nepal, Vol.11, pp. 42-43.

Singh, B. K. (2004), 'Complementary pro-poor programme in community and leasehold forestry', Paper presented in 4th National Workshop on community forestry, DoF, HMG, Kathmandu, Nepal.

Singh, C. (1986),'*Common property and Common Poverty: India's Forests, Forest Dwellers and the bow*', Oxford University Press, New Delhi.

Sunderlin, W., A. Angelsen and S. Wunder (2003), 'Forests and Poverty Alleviation', in FAO, *State of the World's Forests -2003*, Food and Agriculture Organization of the United Nations, Rome, Italy, pp. 61-73.

Thompson, J. (1995), 'Participatory Approaches in Government Bureaucracies: Facilitating the Process of Institutional Change', *World Development*, 23(9): 1521-1534.

Townson, I. M. (1995),' Forest Products and Household Income: A Review and Annotated Bibliography', *Tropical Forestry*, Papers *(31)*, Center for International Forestry Research and Oxford Forestry Institute, Oxford, UK.

Varalakshmi, V. and O. N. Kaul (1999), 'Non-Governmental Organisations: Their role in Forestry Research and Extension', *The Indian Forester, India*, 125(1), pp. 37-44.

World Bank (2000), Towards *a Revised Forest Strategy for the World Bank Group*, The World Bank, Washington, D.C.

World Bank (2001), Recommended *Revisions to OP 4.36*, Proposals for Discussion, The World Bank, Washington, D.C.

31

ECONOMIC EVALUATION OF MULTISTRATA AGROFORESTRY SYSTEM PRACTICED BY TRADITIONAL CULTIVATORS IN NORTHERN BANGLADESH

Syed Ajijur Rahman and Khandaker Mursheda Farhana

INTRODUCTION

Unsustainable exploitation of forest resources and deforestation, decreasing productivity and environmental degradation in the agricultural lands with multiple natural resource management problems are the major environmental challenges for Bangladesh. In recent years one of the most common proposed strategies for addressing environmental degradation in rural areas of the tropics is agroforestry. Agroforestry is an agricultural practice that spans centuries, one that has been used by many indigenous peoples as a traditional land use option, providing sustenance for early agriculturists while preserving forest resources and biodiversity. These early forms of agroforestry have since evolved to be a dynamic ecologically based, institutionalized natural resources management system, broadly defined as the integration of trees on farms or agricultural landscapes. Agroforestry diversifies and intensifies production; both for subsistence and cash, while maintaining forest cover and associated biodiversity (Nair, 1989).

The local farmers in northern Bangladesh have practiced a traditional monoculture for decades and depend on it for their livelihood. Regardless of opinions on the merits or demerits of traditional cultivation, some farmers have recognized the need for modifying the traditional farming practices and the value of conserving soil resources for economic growth and poverty reduction. Some alternative farming technologies (e.g., fruit gardening, alley cropping, multipurpose homegardens) have been introduced in northern Bangladesh. Multistrata agroforestry system is one of the most important and modern cultivation practices of these farming technologies. Because, it can potentially deliver a sequence of environmental, economic, and social benefits, and this system is taken as the point of concentration of the present paper.

MATERIALS AND METHODS

This research focused on Capasia village of Paba Upazila (Upazila is a small administrative unit) under Rajshahi District in Northern Bangladesh. More exactly study area is lies between 24°252″ to 24°202″ north latitudes and between 88°402″ to 88°452″ east longitudes.

Figure 1
Hydro Meteorological Characteristics of the Study Site

Climate	Tropical monsoon
Mean temperature	Maximum - 32^{0}c to 36^{0}c Minimum - 07^{0}c to 16^{0}c
Average humidity	Dry season - 45 to 71 per cent Rainy season - 84 to 92 per cent
Average rainfall	1,448 mm
Topography	Flood plain
Soil condition	Loamy
Average elevation	60 ft.

Research methods used were (1) structured interviews of 60 households and 6 indepth case study focused especially on experiences and capacities; actual and envisaged costs and benefits of agroforestry and monoculture, (2) FGD and observation were also carried out in order to determine actual species combinations, and envisaged costs and benefits of both types of land use. Other data was gathered by way of interviews with key informants (Government, non-government and public organizations) and market prospecting. The secondary data is used from statistical yearbooks, local administrative and various related sources. Research data was collected from the period of May to October, 2006.

For many elements of the study (the model of multistrata agroforestry, actual and envisaged cost and benefit etc.), semi-quantitative analysis was carried out by the use of econometric techniques, especially cost-benefit analysis of multistrata agroforestry and traditional monoculture at the household level. The *net present value* (NPV), *internal rate of return* (IRR), *benefit-cost ratio* (B/C) and *pay back period* of multistrata agroforestry and monoculture were calculated and compared. The NPV determines the present value of net benefits by discounting the streams of benefits and costs back to the beginning of the base year.

The NPV is calculated by the following formula:

$$\sum_{t=1}^{n} \frac{(B_t - C_t)}{(1+r)^t}$$

where

B_t – the benefits of production by a cultivation practice.

C_t – the costs of production by a cultivation practice.

t – the year time.

r – the discount rate.

The IRR is equal to the discount rate (r) that brings the NPV down to zero. An investment is considered financially attractive if the IRR is higher than the opportunity cost of project finance (i.e. what one would pay to the bank for borrowing the investment capital). Following the definition, IRR is obtained by solving the equation:

$$\sum_{t=0}^{n} \frac{B_t - C_t}{(1+r)^t} = 0$$

The B/C compares the discounted benefits to discounted costs. A B/C of greater than 1 means the project is profitable, whilst a B/C of less than 1 means the project generates losses. The B/C is calculated as follows:

$$\frac{B}{C} = \frac{\sum_{t=0}^{n} \frac{B}{(1+r)}}{\sum_{t=0}^{n} \frac{C_t}{(1+r)^t}}$$

The *pay back period* measures the number of years it will take for the undiscounted net benefits to repay the investment. An investment can be acceptable only if its benefits can offset all investment costs within a limited arbitrary period.

The *sensitivity analysis* also carried out to study the effects of the change in mutable factors such as input and output prices, yields of products and discount rate on farmers' benefit. In *sensitivity analysis* the results of economic analysis are checked by considering the effects of changes in the value of key variables.

MULTISTRATA AGROFORESTRY

The term "multistrata" describes agroforestry systems that incorporate combinations of woody plants, such as timber and/or fruit trees, along common agricultural crops, in a same area and at the same time, so that multiple layers of vegetation form. Crops are usually the dominant product while the trees get established and grow (Schroth *et al.*, 2001). Multistrata agroforestry systems can be very labor and knowledge intensive, and thus can require both commitment on the part of the farmer and good initial training and technical support. Table 1 adds further description of multistrata systems along with other common agroforestry systems. Thus, with reference to Table 1, the agroforestry systems examined in this paper correspond to the last two multistrata categories, multi-purpose trees of croplands and plantation crop combinations.

Table 1
Attributes of Select Agro-forestry Systems

Agroforestry system	*Brief description*	*Components (W = woody H = herbaceous)*	*Primary role of woody components (Prt = Protective Prd = Productive)*	*Agro-ecological adaptability*
Improved or enriched fallow	Woody species planted and left to grow during the "fallow phase."	W: fast growing, preferable leguminous H: common agricultural crops	Prt: soil fertility and stability Prd: wood products	In shifting cultivation areas
Taungya	Combined stand of woody and agricultural species during early stages of establishment of plantations	W: usually plantation forestry species, i.e. (*Swietenia macrophylla*) H: common agricultural crops	Prd: additional income from forestry species	In most ecological regions; several improvements possible
Multi-layer tree gardens (multi-strata)	Multi-species, multi-layer, dense plant associations	W: different woody components of varying form H: usually absent	Prt: soil conservation; efficient nutrient cycling Prd: various	Areas with fertile soils, good availability of labor, and high human population pressure
Multi-purpose trees of crop lands (multi-strata)	Trees scattered or arranged according to some pattern within boundaries	W: multi-purpose trees and other fruit trees H: common agricultural crops	Prt: fencing, social values, plot demarcation Prd: various tree products	In all ecological regions, esp. in subsistence farming. Sometimes integrated w/animals.
Plantation crop combinations (multi-strata)	Integrated multi-story mixtures of plantation crops, arranged in some pattern, with possibly some shade trees and other crops	W: combination of crops i.e. coffee, coconut, or other fruit trees, or forestry species H: common agricultural crops usually present, especially with intercropping arrangements	Prt: shade, windbreak, soil protection Prd: large number of products	In humid, sub-humid regions (depending on adaptability of plantation crops); usually in small-holder subsistence system.

Source: Adapted from Nair, 1989

SELECTED AGROFORESTRY MODEL IN THE STUDY SITE

The multistrata agroforestry system in the study area is chosen since it is an important agroforestry model that is appropriate for the cultivation fields, and mostly popular and widely accepted by traditional cultivators. The litchi (*Litchi chinensis*) is planted as a multipurpose tree species in this model. Crop species, i.e., ginger (*Zingiber officinale*), turmeric (*Curcuma domestica*), brinjal (*Solanum melongena*), lemon (*Citrus limonum*), papaya (*Carica papaya*) and banana (*Musa species*) are intercropped with litchi as crop species. In spite of the very small average size of the management units, multistrata agroforestry system is characterized by high species diversity and usually 3-4 vertical canopy strata, which result in intimate plant associations. The lower strata is usually be partitioned into two, with the lowermost (less than 1 m height) dominated by brinjal, ginger and turmeric; the second strata (1-3 m height) is composed by banana, papaya and lemon. The upper stratum is dominated by litchi, which is continuing to grow taller. The layered structure is never static; the pool of replacement species results in a productive structure, which is always dynamic while the overall structure and function of the system are maintained. Brinjal, papaya and banana are intercropped in the beginning years (first 4 years). Lemon is cultivated for the first 6 years. After that time the shade-tolerant ginger and turmeric are commonly planted under litchi trees with the purpose of income generation and rational utilization of sunlight and improvement of soil protection and erosion control.

ECONOMIC ANALYSIS OF MULTISTRATA AGROFORESTRY AND MONOCULTURE

The analysis of collected data is to estimate the costs and benefits from multistrata agroforestry and traditional monoculture. The actual estimation of agroforestry and monoculture vary widely, and depend upon site conditions as well as on many factors such as tree spacing, maintenance techniques, and the rate of growth of trees. For the analysis, all values are based on August 2006 prices, which are held constant throughout the term of the analysis. Analysis is given consistently on a $ (taka)/acre basis.

Assumptions

Land

The real estate market is underdeveloped in the study area, especially for land under traditional cultivation. Therefore, the price of land is difficult to identify. However, as mentioned in MacDicken (1990), there is no need to value the land if farmers want to change existing land use to agroforestry because it would be canceled out in a "with" and "without" comparison. Thus, for simplicity and with the assumption that the value of land is the same and does not change over time for both cultivation practices, it is neglected from the calculation.

Establishment Cost

Establishment cost involves land preparation, seedlings, planting etc. Costs for preparation vary greatly depending on the condition of the site. The farmers in the study area purchased seeds and seedlings from private or state sources, and price ranges vary depending on species and quantity grown. The average establishment cost is tk. 40,000.00 for multistrata agroforestry and tk. 30,000.00 for monoculture per acre (Table 2).

Table 2
Assumptions of Multistrata Agro-forestry and Monoculture

	Operation	*Year*	*Approximate worth (Taka /acre)* Multistrata agroforestry	Monoculture	*Comments*
	Establishment cost	–	40,000.00	30,000.00	This cost is required for the beginning year of an agro-forestry and monoculture.
	Labor cost	1-10	4,000.00	1,500.00	Average yearly labor costs including weeding, thinn-ing, pruning, harvesting, protection etc.
		11-30	5,000.00	1,500.00	For agroforestry, year 11 to 30 litchi gives high yields, which required high protection and harvesting cost.
	Pesticide cost	1-30	–	300.00	Pesticide is only used for monoculture, and it is a common picture of the study area.
	Fertilizer cost	1 -6	6,000.00	3,000.00	For agroforestry: during the
		7-10	5,000.00	3,000.00	time period tree species growing up, it provides
		11-30	3,000.00	3,000.00	nutrients so fertilizer is gradually less used.
	Irrigation	1-30	15,000.00	3,500.00	Irrigation cost of agroforestry is high because of on farm species diversity and shortening fallow period.
Yields	Rice and wheat	1-30	–	45,000.00	All costs are calculated
	Brinjal	1-4	8,000.00	–	average yearly basis.
	Papaya	1-4	50,000.00		
	Banana	1-4	30,000.00		
	Lemon	1st	15,000.00		
		2nd	45,000.00		
		3-6	1,00,000.00		
	Ginger and turmeric	7-30	60,000.00		
	Litchi	4-6	30,000.00		
		7-10	50,000.00		
		11-30	2,50,000.00		

Note: US$ 1 = Taka 68 as of August 2006.

Labor

Family labor is mainly used for farm work in the study site. Yet, hired laborers are also accepted in the area. However, family labor is not cash expenditure from the farmer's perspective. Therefore, all calculations will be carried out for two variants: with and without inclusion of family labor in the production costs. The scenarios with the exclusion of family labor costs seem more meaningful for poor farmers. The labor costs in the study area is tk. 100/workday, which is the common price in this region.

Pesticide and Fertilizer

Pesticide and fertilizer are often used for enhance the farm production. The average cost of pesticide and fertilizer of monoculture is same from beginning to last year. But, for multistrata agroforestry the fertilizer is gradually less used, because the tree species can provide nutrients for understory crops. On the other hand the pesticide is not used for agroforestry and it is common picture of the study area.

Irrigation

Irrigation is essential during the dry season. In the study area average yearly irrigation cost is calculated 15,000.00 for agroforestry and 3,500.00 for monoculture.

Discount Rate

The analysis is carried out from the farmer's perspective, thus the discount rate applied here is not social, but individual. It can be referred to as the opportunity cost of capital for farmers, which is closely related to the financial interest rate. Therefore, the real discount rate of 10% can be obtained by deducting the inflation rate of 4% from the average nominal interest rate of the loans for agricultural production (14%) in the banking system. However, to have a better look at the problem, a sensitivity analysis will be carried out for a range of different discount rates.

Yields

For calculation of yields, two common and popular cash earning crops i.e., rice and wheat are considered for traditional agriculture. And for multistrata agroforestry litchi, ginger, turmeric, brinjal, lemon, papaya and banana are considered. Brinjal, papaya and banana are intercropped with litchi only for the first four years, and lemon is for first 6 years. After that time shade tolerant ginger and turmeric are intercropped. Yields of litchi are calculated under three categories, 1) fourth to sixth year low yields, 2) seventh to tenth year medium yields, 3) eleventh to thirtieth year high yields.

Time Horizon for Analysis

Once forest trees are included in the agroforestry system, its life can be considered infinite. However, the productive life of the model is 30 years and it is also taken as the time horizon for present analysis.

Results

Research data indicates that multistrata agroforestry is more beneficial than traditional monoculture[1]. The financial picture of the two options shows that, from initial year, the return from multistrata agroforestry is gradually increases but the average return of monoculture is constant (Figure 2). The average yearly income of multistrata agroforestry is tk. 2,27,300.00 substantially higher than traditional monoculture (Table 3).

Figure 2
Net Returns of Multistrata Agro-forestry and Monoculture

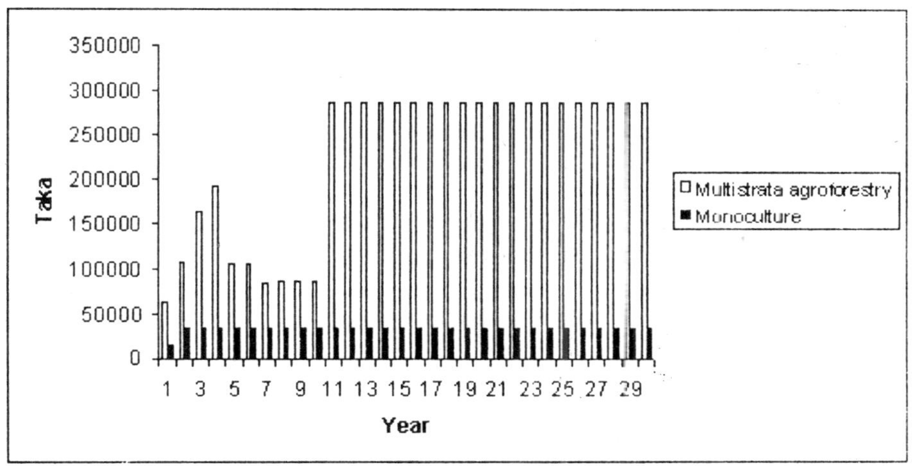

Table 3
Statistical Analysis of Multistrata Agro-forestry and Monoculture (Taka /acre)

Indicators	*Multistrata agroforestry*	*Monoculture*
Mean	2,27,300.00	34,333.33
Median	2,87,000.00	35,000.00
Std. Deviation	88,622.08	3,651.48
Minimum	63,000	15,000
Maximum	2,87,000	35,000

The calculations of NPV, IRR, B/C and payback period of monoculture and multistrata agroforestry systems are illustrated in Table 4. In both scenarios with and without the inclusion of family labor costs, the multistrata agroforestry system gives positive and much higher NPVs than monoculture. The B/C of multistrata agroforestry is more than twice that of monoculture in the both scenario with and without the inclusion of family labor costs. Moreover, the IRR of multistrata agroforestry is also high, so the calculation clearly shows that multistrata agroforestry is more attractive, as it provides a higher yearly net cash flow than monoculture.

Table 4
Financial Results of Multistrata Agro-forestry and Monoculture (Taka /acre)

	Costs including family labor		*Costs not including family labor*	
	Multistrata agroforestry	*Traditional agriculture*	*Multistrata agroforestry*	*Traditional agriculture*
NPV (at 10%) (taka)	1,620,331.73	311,760.19	1,666,425.79	331,809.65
IRR (at 10%)	208%	86%	289%	131%
B/C	7.70	3.77	9.51	4.59
Payback period	1 year	2 years	1 year	1 year

Sensitivity Analysis

Prices can fluctuate significantly over a long time period, especially if overall output increases due to increases in traditional agriculture or agroforestry system adoption, thus reducing the farmers' benefits. Farmers' benefits are also reduced if the opportunity costs of resource use increases. Therefore, sensitivity analysis is considered for the reduction of yields and increase in discount rate. All sensitivity analysis is shown up in Tables 5 to 8.

In relation to discount rate, the NPV of agroforestry systems are positive and higher than NPV of monoculture with the rate up to 100% for both scenarios with and without the inclusion of family labor costs . Normally, the discount rate is not likely to go up this high and hence, multistrata agroforestry systems are generally more profitable than monoculture. In the case of decrease in yields of annual crops, the NPV of agroforestry systems are also found to always be positive and higher than monoculture. The diversification of planted species contributes to the reduction of the risk of output losses for agroforestry system.

Table 5
Sensitivity Analysis of Monoculture Including Family Labor Costs

Sensitivity to decrease in yields		*Sensitivity to change in discount rate*	
Decrease in yields	*NPV*	*Discount rate*	*NPV*
0%	311,760.19	5%	518,988.17
1%	308,642.59	10%	311,760.19
2%	305,524.98	20%	157,596.11
3%	302,407.38	30%	101,237.52
4%	299,289.78	40%	73,210.67
5%	296,172.18	50%	56,666.30
6%	293,054.58	60%	45,833.29
7%	289,936.97	70%	38,235.29
8%	286,819.37	80%	32,638.89
9%	283,701.77	90%	28,362.57
10%	280,584.17	100%	25,000.00

Table 6
Sensitivity Analysis of Monoculture Excluding Family Labor Costs

Sensitivity to decrease in yields		*Sensitivity to change in discount rate*	
Decrease in yields	*NPV*	*Discount rate*	*NPV*
0%	331,809.65	5%	548,237.32
1%	328,491.55	10%	331,809.65
2%	325,173.46	20%	170,481.18
3%	321,855.36	30%	111,235.61
4%	318,537.26	40%	81,603.37
5%	315,219.17	50%	63,999.62
6%	311,901.07	60%	52,395.79
7%	308,582.98	70%	44,201.67
8%	305,264.88	80%	38,125.00
9%	301,946.78	90%	33,450.29
10%	298,628.69	100%	29,750.00

Table 7
Sensitivity Analysis of Multistrata Agro-forestry Including Family Labor Costs

Sensitivity to decrease in yields		*Sensitivity to change in discount rate*	
Decrease in yields	*NPV*	*Discount rate*	*NPV*
0%	1,620,331.73	5%	3,040,189.00
1%	1,604,128.41	10%	1,620,331.73
2%	1,587,925.09	20%	692,125.30
3%	1,571,721.78	30%	411,577.43
4%	1,555,518.46	40%	288,989.30
5%	1,539,315.14	50%	221,463.37
6%	1,523,111.82	60%	178,625.80
7%	1,506,908.51	70%	148,946.35
8%	1,490,705.19	80%	127,158.10
9%	1,474,501.87	90%	110,504.51
10%	1,458,298.56	100%	97,391.11

Table 8
Sensitivity Analysis of Multistrata Agro-forestry Excluding Family Labor Costs

Sensitivity to decrease in yields		*Sensitivity to change in discount rate*	
Decrease in yields	*NPV*	*Discount rate*	*NPV*
0%	1,666,425.79	5%	3,114,459.13
1%	1,649,761.53	10%	1,666,425.79
2%	1,633,097.27	20%	717,692.92
3%	1,616,433.01	30%	429,705.78
4%	1,599,768.75	40%	303,336.23
5%	1,583,104.50	50%	233,486.44
6%	1,566,440.24	60%	189,051.93
7%	1,549,775.98	70%	158,194.21
8%	1,533,111.72	80%	135,493.38
9%	1,516,447.46	90%	118,107.81
10%	1,499,783.21	100%	104,391.60

DISCUSSIONS

Derived from above results, it is clear that multistrata agroforestry is financially more benefited than monoculture. Experimental evidence supporting claims of beneficial effects of multistrata agroforestry is provided by a number of studies. Research (Rahman and Islam, 1997) conducted in Dinajpur, Bangladesh indicated the NPV of agroforestry is positive under all three situations i.e., forest department situation, participant situation and whole agroforestry system situation. The B/C ratios stand at 1.43, 2.21 and 1.95 respectively in all situations. The IRR under the above three situations are 21%, 250% and 42% respectively. Cashew plantations established at central Tamil Nadu, India under an agroforestry system also showed comparable financial results giving B/C ratio of 1.65 and IRR of 40.83% (Sekar and Karunacharan, 1994). Elevitch and Wilkinson (2000) also reported the cost and benefit of agroforestry system in Hawaii is financially viable and less risky than traditional agriculture and forestry.

Agroforestry is not only economically promising but also environmentally sound. Inside agroforestry both tree and crop species play a critical role for agroclimate. It can improve soil fertility by providing nutrients, conserve water, improve soil moisture and protect it from erosion as well as improving biodiversity (Elevitch and Wilkinson, 1998). Agroforestry system is most extensive in developing countries where approximately 1.2 billion poor people depend directly on a variety of agroforestry products and services (IPCC, 2000). In the five sub-Sahara African case studies in Franzel and Scherr (2002), agroforestry is shown to have potential to increase farm incomes and solve difficult environmental problems. It is financially more profitable to local farmers in comparison with traditional cultivation, beside its other economic and social benefits. Thus, it can be a potential alternative cultivation practice that helps to enhance poverty reduction and transition to permanent cultivation (Mai, 1999).

CONCLUSIONS

Multistrata agroforestry exemplify many agroforestry characteristics, i.e., the intimate mix of diversified agricultural crops and multipurpose trees fulfill most of the basic needs of traditional cultivators and local population while the multistoried configuration and high species diversity help to produce sustained yields in a most resource-efficient way. Some traditional farmers in Northern Bangladesh are practicing multistrata agroforestry that is well organized, ecologically sound, and could provide high yields. In the farmers' own perspective, the cost-benefit analysis shows that multistrata agroforestry system is more profitable and less risky than traditional monoculture. On the other hand, monoculture becomes increasingly unsustainable due to the population growth and the loss of environmental conservation methods, which drives farmers into a vicious cycle of poverty and may deteriorate the social and economic life as well. However multistrata agroforestry system requires comparatively higher investment in the beginning years. Whilst, the assessment of multistrata agroforestry system is underestimated since some of environmental benefits are not quantifiable due to the lack of available technical data.

ACKNOWLEDGEMENTS

Authors are thankful to the farmers at the site where field survey was undertaken, who share their precious time, thought and concerns, and even their scarce meals with us. Special thanks also given to the farmers' family for their hospitality and friendship, and most of all their help and support. We also appreciate to Professor Abdur Rahman Siddiquee, Professor Muhammad Hasan Imam and Dr. Abdul Wadud for their valuable advises and guidance.

NOTE

1. Some risk factors such as floods, droughts, and insects are excluded.

REFERENCES

Elevitch C. and Wilkinson K. M. 1998. *A Guide to Orchard Alley Cropping for Fertility, Mulch and Soil Conservation.* AgroForester, Hawaii.

Elevitch C. R. and Wilkinson K. M. 2000. Economics of farm forestry: financial evaluation for landowners. pp. 3-31 in Elevitch C.R. and Wilkinson K. M. (eds). *Agroforestry Guides for Pacific Islands # 7.* Permanent Agricultural Resources (PAR), Hawaii.

Franzel S. and Scherr S. J. 2002. Introduction. pp.1-11 in Franzel S. and Scherr S. J. (eds). *Trees on the Farm: Assessing the Adoption Potential of Agroforestry Practices in Africa.* CABI, Wallingford.

IPCC (Intergovernmental Panel on Climate change, 2000. *Land-use, Land-use Change and Forestry.* Special Report. Cambridge University Press, Cambridge.

MacDicken K. G. and Vergara N. T. 1990. *Agroforestry: Classification and Management.* John Wiley & Sons, New York.

Mai P. T. 1999. *Socio-Economic Analysis of Shifting Cultivation versus Agroforestry System in The Upper Stream of Lower Mekong Watershed in Dak Lak Province.* M. A thesis in Economics of Development. College of Economics, National University-HCMC, Viet Nam and Institute of Social Study (ISS), The Netherlands.

Nair P. K. 1985. Classification of agroforestry systems. *Agroforestry Systems,* 3: 92-128.

Nair P. K. 1989. *Agroforestry Systems in the Tropics.* Kluwer Academic Publishers, Dordrecht, Netherlands.

Rahman A. and Islam S. S. 1997. Financial viability of agroforestry under participatory approach in Bangladesh: the case of forest department's first logged over plots of Dinajpur. *Bangladesh Journal of Forest Science,* 26(1): 47-55.

Schroth G., Lehmann J., Rodrigues M. L., Barros E. and Macedo J. L. 2001. Plant-soil interactions in multistrata agroforestry in the humid tropics. *Agroforestry Systems,* 53: 85-102.

Sekar C. and Karunacharan K. R. 1994. Economic analysis of cashew plantations under agroforestry conditions of central Tamil Nadu. *Journal of Tropical Forest Science,* 6(4): 523-528.

32

COST-BENEFIT ANALYSIS OF AGRICULTURE VERSUS AGROFORESTRY SYSTEM IN THE HIGH BARIND TRACT OF NORTHERN BANGLADESH

Syed Ajijur Rahman and Khandaker Mursheda Farhana

INTRODUCTION

The alarming poverty situation and steady population growth in Bangladesh have raised the pressure on natural resources throughout the country[1]. By the year 2003, only 10.2 per cent of the country's surface area was left under forest that, in turn, is being depleted at a rate of 3.1 per cent per annum. Forests are depleted by commercial timber exploitation and gradually converted into pastures, plantations and cultivated fields. The major cause of drastic deforestation is mainly agriculture, which is closely linked to poverty and environmental problems. In poor rural areas, for example, there is a close link among poverty, high fertility, high population growth and extensive deforestation, as peasants fell tropical forests for firewood and new farmland. At the same time, many environmental problems stem from poverty- often contributing to a downward spiral in which poverty exacerbates environmental degradation and environmental degradation exacerbates poverty. In this way, with increasing aridity and recurrent drought, the 74,200-sq km Barind Tract area in the north region of the country is heading towards serious desertification problems (Hossain, 1995). Moreover, due to the loss of firewood resources, rural communities turn to alternative fuels such as cow dung and crop residues that previously served as organic manure on agricultural fields. The shorter fallow periods, reduced organic inputs and repeated removal of nutrients with harvested crops have contributed to soil degradation and, consequently, lowered crop yields and farmers' returns leading to more poverty.

Analysing this situation, in vast rural areas, agroforestry is seen as one of the major strategies for poverty reduction in Bangladesh, where there is little scope of developing pure forest due to obvious priority for food crop production. Because, agroforestry project is not only helps to increase food and fodder but also protects the existing forest as well as improving biodiversity where unemployed and poor people rush to earn their livelihood. In Bangladesh, agroforestry has been practicing in different areas for a long time but got official recognition in1977-78 with the initiation of projects on agroforestry at Betagi and Pomra in Chittagong (Ministry of Agriculture, 1987; Alim, 1988; Rahman, 1991). Crop species cultivates with tree species under agroforestry projects were found successful in various parts of Bangladesh. However, the rates of return of agroforestry is positively significant and much higher compared to many agricultural projects.

OBJECTIVES AND METHODOLOGY

The main objectives of the present paper are to assess the benefit-cost of agroforestry and to prescribe some policy suggestions on how agroforestry can best be applied. This research focused on four different areas of high Barind Tract i.e., Niamatpur, Godagari, Tanore and Nachole under three different districts i.e., Rajshahi, Nawgoan and Chapai Nawabgong in Northern Bangladesh. Research methods were used (1) a census and Rapid Rural Appraisal (RRA) for the basic social circumstances of the areas; (2) short structured interviews with a purposive sample of 200 households focused especially towards experiences and actual and envisaged costs and benefits of agriculture and agroforestry, (3) Participatory Rural Appraisal (PRA) and field-based surveys implemented by the 60 interviewees. Other data was gathered by way of interviews with key informants (GO, NGO, firms and PO) and market prospecting. Moreover, secondary data was gathered and in the form of the CBAs poverty reduction assessment. Research data was collected from the period of January to November, 2004.

Both quantitative and qualitative method were used for analyze data. Cost-benefit analysis is applied to assess the quantitative data. The *net present value* (NPV), *internal rate of return* (IRR), *benefit-cost ratio* (B/C) and *pay back period* of agroforestry and agriculture were calculated and compared.

The NPV is calculated by the following formula:

$$NPV = \sum_{t=1}^{n} \frac{(B_t - C_t)}{(1+r)^t} \tag{1}$$

The IRR can be obtained by solving the following equation:

$$IRR = \sum_{t=0} \frac{B_t - C_t}{(1+r)^t} \tag{2}$$

The B/C is calculated as follows:

$$\frac{B}{C} = \frac{\sum_{t=0}^{n} \frac{B}{(1+r)}}{\sum_{t=0}^{n} \frac{C_t}{(1+r)^t}} \tag{3}$$

The *pay back period* measures the number of years it will take for the undiscounted net benefits to repay the investment. An investment can be acceptable only if its benefits can offset all investment costs within a limited arbitrary period.

DEFINITIONS

Agriculture

Generally agriculture is a science or art of cultivating the soil, growing and harvesting crops, and raising livestock. The most widely accepted definition of agriculture is: The science or business of raising plants and animals useful to man. It implies the cultivation of the soil, the production and harvesting of crops, the care and breeding of livestock (HELCOM PITF, 2001). Agriculture is crop and animal husbandry for producing food stuffs, raw materials and services (including biodiversity and landscape) for the society.

Agroforestry

There are variations of the definition of agroforestry system, but the most widely accepted definition is of the International Center for Research in Agroforestry (ICRAF), which was cited by Nair (1990): Agroforestry is a collective name for land-use systems and technologies where woody perennials are deliberately used on the same land-management units as agricultural crops and/or animals, in some form of spatial arrangement or temporal sequence. In agroforestry systems there are both ecological and economical interactions between the different components.

There are two main components in this definition. The first is the deliberate integration of trees with agricultural crops and/or animals on the same piece of land, and the second is the ecological and economical interactions between woody and non-woody components. The systems that lack of one from these two components cannot be classified as agroforestry (Gregesen *et al.*, 1989).

COST-BENEFIT ANALYSIS OF AGRICULTURE VS AGROFORESTRY SYSTEM

The analysis of collected data is to estimate the costs and returns from agroforestry and agriculture. The actual estimation of agroforestry and agriculture are very widely, and depend upon site conditions as well as on many factors such as tree spacing, maintenance techniques, and the rate of growth of trees. For the analysis, all values are based on January 2005 price, which is held constant throughout the tern of the analysis. Analysis should be given consistently on a $/acre basis.

Selection of Models for Analysis

The alley cropping model of agroforestry in the high Barind Tract area is chosen since it is the important agroforestry model applied in this region that is appropriate for the cultivation fields. Furthermore, this model can be expanded to the surrounding villages. The mango (*Mangifera indica*) trees are planted as a multipurpose tree species in this project. Crop species, i.e., paddy (*Oryza sativa*), wheat (*Triticum aestivum*), ginger (*Zingiber officinale*) and turmeric (*Curcuma domestica*) are intercropped with mango trees for the purpose of income rising. Paddy and wheat are intercropped in the beginning years (first 12 years). After that time the shade-tolerant ginger and turmeric are commonly planted under mango trees with the purpose of income raising and rational utilization of sunlight and improvement of soil protection and erosion control. For the analysis of agricultural project, the monoculture of rice and wheat are considered because these are the most valuable and popular species in this cultivation practice.

The average temperature in this area is 37.8°c for summer and 9.2°c for winter. The average rainfall is over 1300 millimeters per year. Topography is undulating and elevation is 40m from sea level (Table 1).

Table 1
Site Characteristics

Characteristics	*Description*
Climate	Tropical
Topography	Undulating
Slop	5 – 20%
Elevation	40 m.
Rainfall	Over 1300 millimeters per year
Average temperature	Summer 37.8°c, Winter 9.2°c (average)
Dry season	December-march
Soil condition	Silt loam to clay loam
Water resources for irrigation	Mainly ground water, Ponds and natural rainfall

Assumptions

Establishment Cost

Establishment cost involves land preparation, seedling, planting etc. Costs for preparation are very greatly depending on the condition of the site. The farmers of study area were purchased seedling from private or state sources, and range in price depending on species and quantity grown. The average establishment cost is tk. 36,500.00 for agroforestry and tk. 30,000.00 for agriculture per acre (Table 2). Usually land costs are not included in the projected costs as it is assumed that landowner already owns land for this purpose. However, as mentioned in MacDicken (1990), there is no need to value the land if farmers want to change existing land use to an agroforestry or agriculture land use because it would be canceled out in a "with" and "without" comparison.

Table 2
Assumptions of Agriculture and Agroforestry Projects

		Approximate costs		
Operation	*Year*	*Agroforestry*	*Agriculture*	*Comments*
Establishment cost	–	36,500.00	30,000.00	–
Fertilizer	1-9	3,000.00	3,000.00	Years 10 to 30-fertilizer cost (per year) of agroforestry are low than agriculture because in this period tree species provide nutrients.
	10-30	2,500.00	3,000.00	
Pest	1-15	300.00	300.00	Years 16 to 30 pest cost (per year) of agroforestry are higher because large amount of pest is used for fruits protection.
	16-30	600.00	300.00	
Irrigation	Annual	3000.00	3000.00	–
Labor cost	1-15	32,00.00	32,00.00	Cost included wedding, thinning, pruning, harvesting, protection etc. In agroforestry, years 16 to 30 shed tolerate ginger and turmeric are intercropped, that required relatively less amount of labor.
	16-30	2,500.00	32,00.00	

Note: Time horizon for analysis is 30 years. USD 1 = Taka 62 as of January 2005

Yields

For calculation of yields, two common and most popular cash earning crops i.e., rice, and wheat are considered for agriculture. And for agroforestry project rice, wheat, ginger, turmeric and mango are considered. Rice and wheat are considered only for first twelve years after that time ginger and turmeric are measured, because shed tolerate ginger and turmeric are intercropped with mango trees in this time. Mango is calculated under three categories i.e., 1) seventh to eleventh years low yields, 2) twelfth to eighteenth year medium yields and 3) nineteenth to thirteenth years high yields.

Labours

Mainly family labours in the study site are used for farm works. A cultivation practice that requires hired labours is not likely to be accepted in the area. Meanwhile, family labours are not cash expenditure in farmers' perspective. Therefore, all calculations will be carried out for two variants: with and without inclusion of family labour in the production costs. The scenarios with the exclusion of family labour costs seem more meaningful for poor farmers. The opportunity costs of labour used in the case of family labour costs included is tk. 50/workday, which is the common price of hired labour in this region.

Pest and Fertilizer

Pest and fertilizer are often used for enhance the trees and crops. The average cost of pest and fertilizer of agriculture is same from beginning to last year. But, for agroforestry project the fertilizer is less use since tenth year, because in this time tree species can provide nutrients for understory crops. On the other hand the pest demand of agroforestry is increased form that time when trees provide fruits.

Irrigation

Irrigation is needed only for dry season in both of agriculture and agroforestry project. During the dry season in the study area the main source of irrigation is ground water. The average yearly irrigation cost is tk. 1000.00 for both agriculture and agroforestry project.

Discount Rate

The analysis is carried out for farmers' perspective, thus the discount rate applied here is not social, but individual. It can be referred to as the opportunity cost of capital for farmers, which is closely related to the financial interest rate. Therefore, the real discount rate of 10% can be obtained by deducting the inflation rate of 4% from the average nominal interest rate of the loans for agricultural production (14%) in the banking system. However, to have a better look at the problem, the sensitivity analysis will be carried out for a range of different discount rates.

FINANCIAL RESULTS

Research data indicates agroforestry is more benefited than normal agricultural system. The financial picture of two options such as agroforestry and agriculture show that, in the initial year the net return of agriculture is higher than agroforestry. Because establishment cost of agriculture is comparatively low. But from second to last year the average return of agriculture is constant[2]. Hence, the return of agroforestry is gradually higher because from seventh years the tree species also start to give yields (Figure 1).

Figure 1
Net Returns of Agriculture versus Agroforestry Projects

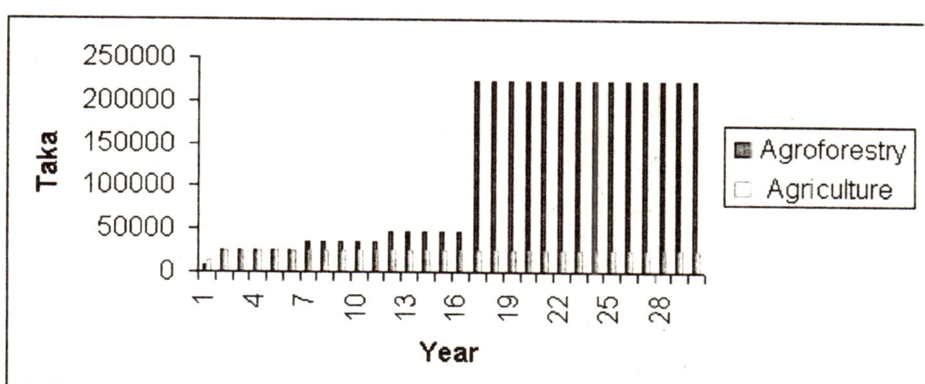

The statistical indicators show that the average yearly income of agroforestry is tk. 1,21,680.00 that is higher than agriculture (Table 3). The minimum income of agroforestry is tk. 7,400.00 and maximum is tk. 2,22,000.00. On the other hand the minimum income of agriculture is tk. 14,000.00 and maximum is tk. 26,000.00. The income range of agroforestry is higher than agriculture.

Table 3
Statistical Indicators of Agriculture and Agroforestry Projects

Indicators	*Agroforestry*	*Agriculture*
Mean	1,21,680.00	25,600.00
Median	45,700.00	26,000.00
Std. Deviation	95,755.25	2,190.89
Minimum	7,400.00	14,000.00
Maximum	2,22,000.00	26,000.00
Range	2,14,600.00	12,000.00

The calculations of NPV, IRR, B/C and payback period of agriculture and agroforestry model are illustrated in Table 4. In both scenarios with and without the inclusion of family labor costs, the agroforestry system gives positive and much higher NPV than agriculture. The B/C of agroforestry is nearly about twice than agriculture. But the IRR of agriculture is higher than agroforestry because in the beginning years the net return of agriculture is high. The calculation clearly makes that the agroforestry is more attractive, which gives yearly higher positive net cash flow than agriculture.

Table 4
Financial Results of Agroforestry Versus Agriculture

	Costs including family labour		*Costs not including family labour*	
	Agroforestry	*Agriculture*	*Agroforestry*	*Agriculture*
NPV (at 10%) (taka)	559,953.89	204,024.56	596,977.75	239,846.83
IRR (at 10%)	46%	56%	58%	73%
B/C	3.55	1.99	4.25	2.37
Payback period	2 years	1 year	2 years	1 year

SENSITIVITY ANALYSIS

The prices can fluctuate significantly over long time period, especially if overall output increases due to increase in agriculture or agroforestry system adoption, thus reduce the farmers' benefits. Farmers' benefits are also declined if the opportunity costs of resource use increases. Therefore, sensitivity analysis is considered for the reduction of yields and increase in discount rate. All sensitivity analysis is shown up in Tables 5 to 8.

In the relation to discount rate, NPV of agroforestry system is positive and higher than NPV of agriculture with the rate up to 40% for both scenarios with and without the inclusion of family labor costs . Normally, the discount rate is not likely to go up to this stage and hence, agroforestry system is generally more profitable than agriculture. In the case of decrease in yields of annual crops, NPV of agroforestry system is always positive and double than agriculture. The diversification of planted species contributes to the reduction of the risk of output losses for agroforestry system.

Research (Rahman and Islam, 1997) conducted in Dinajpur, Bangladesh indicates the NPV is positive under all three situations i.e., forest department situation, participant situation and whole agroforestry system situation. The B/C ratios stand at 1.43, 2.21 and 1.95 respectively in all situations. The IRR under the above three situations are 21%, 250% and 42% respectively. The rates of return are significant and much high compared to many agricultural projects. The IRR of *Eucalyptus camaldulensis* under agroforestry project in seven different areas of Bangladesh i.e., Dinajpur (rural Charkai), Dinajpur (urban Charkai), Sylhet (Chawtali), Sylhet (Lawachara), Tangail (Charaljani), Mymensingh (Santoshpur) and Cox's Bazar (Chainda) were fund 22.87%, 20.66%, 20.57%, 39.13%, 27.58%, 12.67% and 59.81% respectively. For Dinajpur the corresponding B/C ratios were 2.49, 2.48, 2.37 and 2.21 at tree rotation ages of 10, 12, 14 and 16 years respectively. Similar results were found for plantation of Sylhet, Tangail, Mymensingh and Cox's Bazar (Kibria *et al.*, 1999). Cashew plantations established at central Tamil Nadu, India under agroforestry conditions also showed comparable financial results giving B/C ratio of 1.65 and IRR of 40.83% (Sekar and Karunacharan, 1994). Research also indicates the cost and benefit of agroforestry system in Hawaii is financially viable and less risky than agriculture and forestry (Elevitch and Wilkinson, 2000).

UNQUANTIFIABLE COSTS AND BENEFITS

Due to the lack of systematical technical documentation, some costs and benefits of agroforestry system cannot be quantified and included in the cost-benefit analysis. The intangible costs and benefits from farmers' perspective are mainly environmental effects on farm households' agricultural production.

Soil erosion control

The agroforestry system can bring more benefit to farmers over agriculture due to better soil protection. Soil erosion declines with the increase in vegetative cover. In the early year of the agroforestry system, the intercrop of annual crops helps to raise the cover over land. Tree species are very effective protective barriers against soil erosion. Since fifth year the trees will be closed and play critical role in soil conservation. These soil erosion control measures will increase land productivity through maintaining soil depth and quality (Hien, 1998). If there is available data on the amount of reduction in soil loss by agroforestry cultivation and on nutrient composition of the soil, the increase in nutrient amount due to this reduction can be calculated. The benefit from soil protection then can be estimated by the value of equivalent inorganic fertilizer required for obtaining this increase in nutrient. This approach is called replacement cost technique and has been used in Saigal and Norse (1993).

Effects on Soil Fertility

The agroforestry system can improve the soil fertility by the restriction of soil erosion and the presence of the nitrogen fixing species. Hence, they can improve the land productivity and the crop yields. The estimation of benefits from nitrogen fixation could be made by replacement cost method if rates of nitrogen fixation by species were known. However, these data are not available since there have been few direct measurements of nitrogen fixation by trees due to the shortcomings of nitrogen-fixation measurement methods (Nair, 1990).

Water Retention

The tree species in the agroforestry system play an important role in water conservation. The water retention capability of trees can maintain soil moisture; therefore raise productivity of the land.

WIN-WIN EFFECT

Agroforestry has *'Win-Win'* effect and agriculture has *'Win-lose'* effect if we apply "Four-fold classification model" (Figure 2). Because several researches indicate agroforestry is considering as one of the major strategies for poverty reduction as well as protects the environment through increase the existing forest. Where unemployed and poor people can protect watershed services through agroforestry based solutions that enable the poor to be rewarded for their provision of livelihood; assist them to better adapt to change, and to benefit from emerging carbon markets, through tree cultivation. The whole agroforestry system is financially viable and contributes significantly towards land use efficiency, employment generation, environmental stability and social equity.

Figure 2
Four-fold Classification Model of Human Well-being and Forest Cover

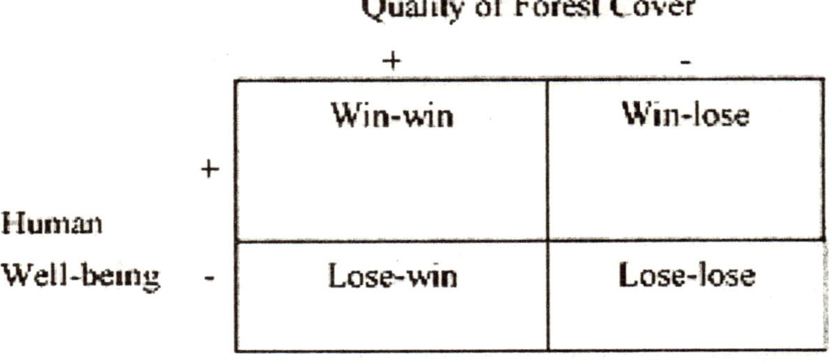

Source: Sunderlin, 2003

This model, applied to the case of forest dwellers and forests, assumes the following definitions:

- *"Win-win"* means poverty alleviation and environmental protection are assumed to go hand-in-hand.
- *"Win-lose"* means poverty alleviation happen at the expense of forests and biodiversity.
- *"Lose-win"* outcomes occur when livelihood security is undermined by excluding local people from forests.
- *"Lose-lose"* outcomes mean that both local people and the environment lose out.

Unsustainable traditional agriculture is the main cause of forest loss and degradation, decrease in soil fertility due to soil erosion and lower productivity. The current trend of drought and desertification problem in high Barind Tract is the results of extend agricultural land, which results in negative impacts on the environment. Poverty is the main motive for unsustainable agriculture and puts into a vicious circle of poverty as shown in Figure 3.

Please Insert Figure 3 here

However, for farmers' subsistence the agriculture still has its advantages over agroforestry system:

- First of all, in the short term when the effects of shortened fallow and land degradation are not yet exposed, agriculture can always give positive and relatively stable net cash flow each year. Meanwhile, agroforestry system causes low returns in beginning years.
- Secondly, the establishment cost of agroforestry is high, and so, agriculture is particularly attractive for poor farmers who are always lack of cash and whose family labors are main and almost the only input for agricultural production.

CONCLUSION AND RECOMMENDATIONS

Some farmers of High Barind Tract in northern Bangladesh are practiced agroforestry that is well organized, ecological sound, and could provide high yields. On the other hand, agriculture becomes increasingly unsustainable due to the shortening of fallow periods and the loss of environmental conservation methods. That drives farmers into a vicious circle of poverty and may deteriorate the social and economical life as well. In farmers' perspective, the financial analysis shows that agroforestry system is more profitable and less risky than agriculture. The NPV of agroforestry system is positive and higher than NPV of agriculture with the rate up to 40% for both scenarios with and without the inclusion of family labor costs. However agroforestry system requires higher investment, generates low returns in the beginning years. The cost and benefit of agroforestry system are underestimated since some of them are not quantifiable due to the lack of available technical data.

The following recommendations can be consider for improving agroforestry system:

- Choice of trees and their provenance suitable to that locality to be encouraged;
- Preference should be given to indigenous and fast growing species;
- Adopt correct spacing recommended to each species with reference to soil and climate;
- Crop combination in agroforestry should be selected with minimum negative interactions;
- Multi-tier system of crop combination could be practiced for effective utilization of land as well as non-food crop production could be balanced;

- Ensure supply of quality seedlings/seeds to farmers by the agencies;
- Young trees require protection from cattle, especially in the fallow season, when the village livestock is let loose to browse agricultural residues and stubbles;
- Imparting training to farmers in crop for better management of farming;
- Manpower of the forestry department should be skilled. Because the functionaries in this department are neither properly trained for this new venture nor their mindset for extension is changed due to which the required thrust cannot be maintained.
- Land tenure problems are also important reason for declining interest in agroforestry. Modern land tenure system and update new land law required;
- High quality research is necessary to innovate new idea and technology in the concerned field.

ACKNOWLEDGEMENT

Authors are grateful to the Department of Sociology and the Institute of Bangladesh Studies (IBS), University of Rajshahi, for providing financial support and the necessary equipment for conducting the experiment. We also appreciate the assistance of Dr. Abdul Wadud for giving fruitful direction.

Figure 3
Unsustainable Agriculture is the Cause of ' Vicious Circle' of Poverty

NOTES

1. The most recent estimates of extreme poverty according to the UNDP report 2004, 36.0 per cent of the country's population lives under the extreme poverty line, compared to 6.6, 13.4 and 34.7 per cent for Sri Lanka,

Pakistan and India respectively (UNDP, 2004). Moreover, with an annual growth rate of 1.9 per cent, the total population of Bangladesh may increase from 137,480 thousands in mid 2004 to 192,218 thousands in 2025 (ESCAP, 2004).

2. Some risk factors such as floods, droughts, and insects are excluded.

REFERENCES

Alim, A. (1988), *Forestry With the People and For the People*, Chittagong University: Bangladesh.

Economic and Social Commission for Asia and the Pacific (ESCAP) (2004), *ESCAP Population Data Sheet 2004*, Population and Rural and Urban Development Division: Bangkok.

Elevitch, C. R. and K. M. Wilkinson (2000), Economics of Farm Forestry: Financial Evaluation for Landowners, In C. R. Elevitch and K. M. Wilkinson, eds, *Agroforestry Guides for Pacific Islands No. 7*, Permanent Agricultural Resources (PAR), Hawaii.

Gregersen, H., S. Draper and D. Elz (1989), People and Trees: The Role of Social Forestry in Sustainable Development, EDI Seminar Series, World Bank: Washington D.C.

Helsinki Commission Programme Implementation Task Force (HELCOM PITF) (2001), 'Investment Activities/Point Source Pollution and Non-Point Source Pollution, 18th Meeting, 15-16 November, Tallinn, Estonia.

Hien,T. Q. (1998), Final Report on the Joint Development of Agroforestry Models in Bu Juk Hamlet, Dak Phoi Commune, Lak District, Dak Lak Province, Unpublished Short-term Mission Report of SUMP, MRC/GTZ: Viet Nam.

Hossain, A. (1995), World Desertification Day: Bangladesh Perspective, In Daily Star, 17 June 1995, Dhaka, Cited in Economic and Social Commission for Asia and the Pacific (ESCAP) (1997), *Population and Environment Dynamics, Poverty and Quality of Life in Countries of the ESCAP Region*, Asian Population Studies Series, No.147, United Nations : New York.

Kibria, M. G., M. A. T. Hossain, D. C. Sarker, M. M. Haque and M. A. Mannan (1999), Financial Analysis of *Eucalyptus Camaldulensis* Plantation in Bangladesh, Bangladesh Journal of Forest Science, 28(2): 102-108.

MacDicken, K. G. and N. T. Vergara (1990), *Agroforestry: Classification and Management*, John Wiley & Sons: New York.

Ministry of Agriculture (1987), Participatory Forestry in Bangladesh: Concepts, Experiences and Recommendations, A Task Force Report of the Ministry of Agriculture, Government of the People's Republic of Bangladesh, Dhaka.

Nair, P. K. R. (1990), The Prospects for Agroforestry in the Tropics, World Bank Technical Paper No. 131, World Bank: Washington D.C.

Rahman, A. (1991), Social Forestry in Bangladesh: The Betagi-Pomra Experience, Paper Presented at the Seminar on Betagi-Pomra Community Forestry Projects Organised by the Bangladesh Agricultural Research Council, December 15: Dhaka.

Rahman, A. and S. S. Islam (1997), Financial Viability of Agroforestry under Participatory Approach in Bangladesh: The Case of Forest Department's First Logged over Plots of Dinajpur, *Bangladesh Journal of Forest Science*, 26(1): 47-55.

Saigal, R. and D. Norse (1993), National Economic Cost of Soil Erosion in Zimbabwe, In M. Munasinghe, ed, Environmental Economics and Natural Resource Management in Developing Countries, CIDIE: Washington D.C.

Sekar, C. and K. R. Karunacharan (1994), Economic Analysis of Cashew Plantations under Agroforestry Conditions of Central Tamil Nadu, *Journal of Tropical Forest Science*, 6(4): 523-528.

Sunderlin, W. D. and T. B. Huynh (2005), *Poverty Alleviation and Forests in Vietnam*, Center for International Forestry Research (CIFOR): Indonesia:.

United Nations Development Programme (UNDP) (2004), *Human Development Report 2004*, Oxford University Press: New York.

33

HUMAN DIMENSIONS IN PROTECTED AREA MANAGEMENT: TRACING SYNERGIES OF CONSERVATION AND LIVELIHOOD

V. Anitha, P. K. Muraleedharan and A. S. Binilkumar

INTRODUCTION

The pressure of a growing population and ever expanding human activities has increased the demand on the scarce land resources. Recent reports by the World Resources Institute have shown that more than 80 per cent of the planet's natural forests have already been destroyed (Hatch, 1997). In the tropical forests of the world, the clearing of the land for agriculture and livestock are the primary activities resulting in deforestation and the main cause is unequal distribution of land (Anderson, 1990). The market and policy changes also act as driving forces of deforestation (Gutman, 2001). Research has shown that people participation is vital in sustainable resource management (Palik, 1993; Poffenberger, 1990; Colfer, 1995). Yet, the implementation of a biologically focused approach to conservation and the existence of pressures from politically powerful vested interest groups, have contributed to the cultural and socio-economic marginalisation of many people living within or close to PAs. These resource use conflicts are widespread and span most of the continents (Anderson, *et al.* 1996, Arild, 2001; Chambers, 1997; Claude and Michelle, 2000; Michael, 2000). All these reveal the intricacies of the human dimensions involved in natural resource management.

Several forest areas have been demarcated as Protected Areas (PAs) due to their ecological significance. PAs perform a large number of social and economic functions also. PAs contribute to the maintenance of ecological processes, species diversity and the genetic variation within species. They safeguard ecosystems, maintain or enhance the productive capacity of the ecosystems, facilitate scientific research and education, contribute ecological functions of the adjoining ecosystems and also help to maintain the cultural capital of the local people (Gadgil, 1992; McCleery,

2000). PAs usually have multiple conservation objectives, such as the conservation of genetic resources, the control of erosion and sedimentation, the maintenance of indigenous use and the provision of recreation and tourism services. PAs do face threats from the high dependencies of the local communities upon the local resources for their sustenance needs because in India these PAs are the only remaining sources of fuel, fodder, other forest produce, fish and other aquatic resources, with most surrounding areas having been degraded (Kutty and Kothari, 2001). These multiple roles of the PAs covering a wide range of environmental and livelihood needs make it important that rational utilisation of forest resources are crucial for human welfare and sustainable socioeconomic development.

In this background, the corpus of this paper is to assess the socioeconomic implications of the local communities dependencies on the natural resource and enumerate the human-related constraints in management as a result of conflicting Protected Area objectives and basic livelihood needs on two Wildlife Sanctuaries in the Kerala State, namely, Peechi-Vazhani and Chimmoni.

STUDY AREA

Peechi-Vazhani and Chimmoni WLSs, located in the Peechi Wildlife Division lie contiguous to each other (Map 1). Peechi-Vazhani Wildlife Sanctuary, located in Thrissur district of Kerala state, lies between 10° 25′ and 10° 40′ N. Latitude and 76° 15′ and 76° 30′ E. Longitude. Total area of sanctuary is 125 km^2. There are two dams in the sanctuary with an area of 12.25 km^2 and 1.843 km^2 respectively (Mammen, 2000; Narayanakutty, 1990). Chimmoni Wildlife Sanctuary is situated east of the Parambikulam Wildlife Sanctuary, Thrissur district, with 10° 22′ N to 10° 26′ N latitudes and 76° 31′ E to 76° 37′ E longitude on the western outer slopes of the *Nelliampathies*. Keeping in view that there exists no significant differences in terms of landscape elements and parameters the socioeconomic assessment of the dependent communities is done for both the sanctuaries together taking them as a single unit.

ANALYTICAL FRAMEWORK AND METHODS

The study was based mainly on primary data, supplemented with secondary data wherever necessary. A multistage stratified random sampling method was adopted for the selection of samples, the unit of study being the household. There are a total of 11 panchayats (Village councils) within which the two sanctuaries lie; among which the study area covered 39 wards hosting approximately 6,000 households in and around the PAs. Taking these households as population, approximately 10 per cent sample i.e., 600 households were selected for the socioeconomic assessment giving more weightage to the tribal households on the fact

Map1: Study Area

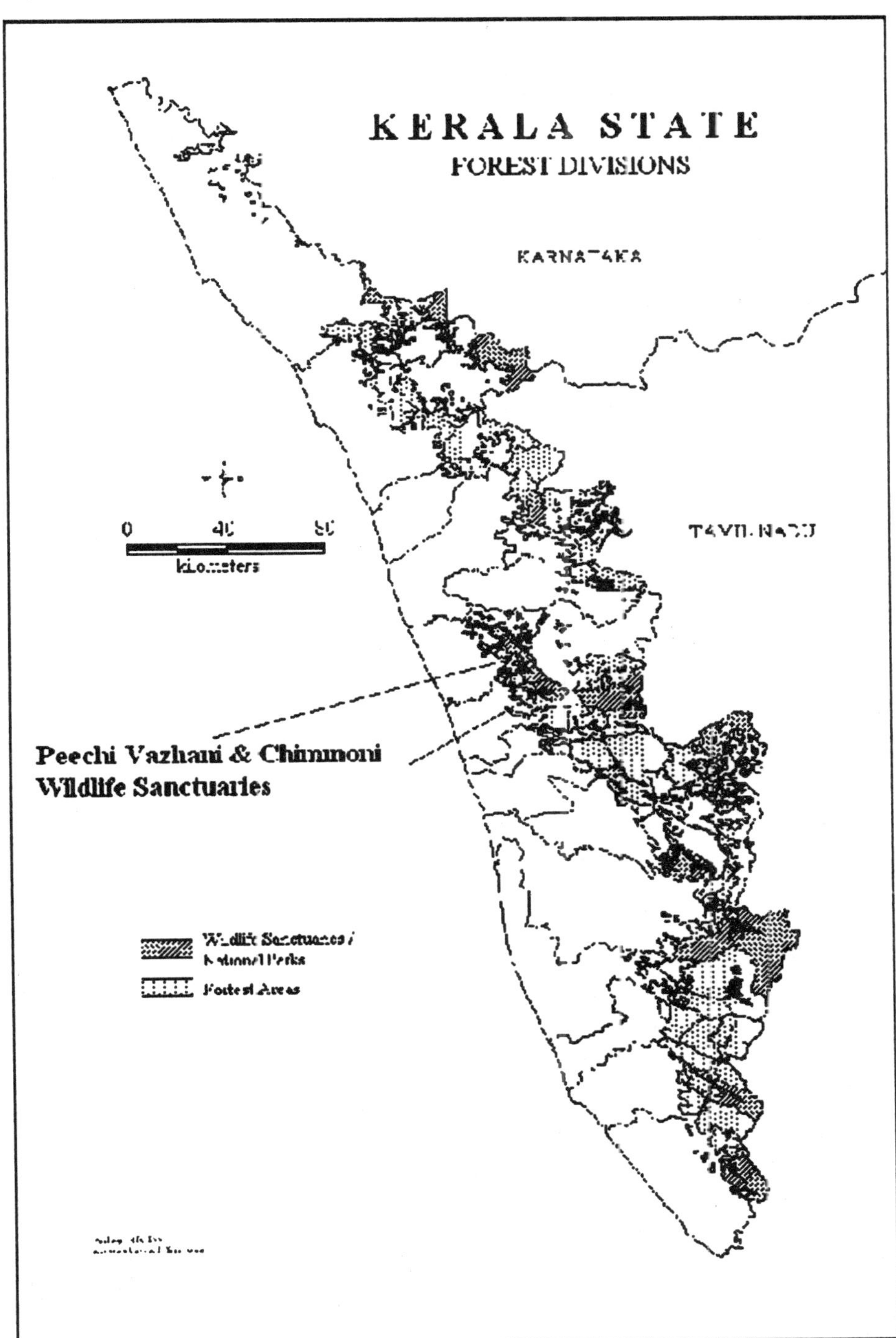

that they are more dependent on the forest resources. There are a total of 25 tribal settlements covering both the WLSs. The criteria for sample selection were proximity (current users), encroachment, tribal settlements, fringe areas (boundaries), enclave settlements and extraction of NTFPs.

To test the existence of association between forest dependency and social group, a chi-square (χ^2) test was employed. The inequality index of the stakeholders has been worked out using the formula

$$T = \sum_{i=1}^{n} (y_i/Y)\log\left(\frac{y_i/Y}{1/n}\right) \tag{1}$$

When there is perfect equality T values become zero. When there is a perfect inequality $T \to \log(n)$. A linear Multiple Regression Analysis was carried out in order to assess the relationship between income from Non-Timber Forest Products (NTFPs) and other socioeconomic parameters. The human related constraints in management were identified based on PRA exercises, informal discussions with local communities and the FD officials and field staff during the study period 2000-2002.

RESULTS AND DISCUSSION

Socioeconomic Profile of Target Group

The survey revealed that all the local stakeholders in Peechi-Vazhani had migrated to their present habitation during the last many decades and is dependent primarily on agriculture. They had cleared the forests during earlier periods when the Government promoted forest clearance for food production. The Forest Policy 1988 does not allow any more clearance of forests for agriculture or other purposes, which has put a halt on the same.

The local community (sample) feature reflects a heterogeneous entity with 44 per cent tribals (*Malayans* and *Kaders);* 33 per cent pioneer rice cultivators, migrant farmers (new settlers who came during the last 40 years); 11 per cent of scheduled castes (*Kavaras, Vettuvas, Pulayas, Parayas*) and other backward communities (Anitha and Muraleedharan, 2002). The occupational pattern reflects primary sector dominance with high dependence on non-forestry income. Majority of the local communities in the study area constitute the low-income segment of the society. The annual household income of majority of the local community (51.33 %) fall in the range of Rs.10,000 – 25,000, all of whom belong to socially and economically weaker sections of the society. The mean income of the tribal households and non-tribal households were calculated separately in order to understand the real picture of the income of the tribes and non-tribes, which showed significant difference between the two groups (Table 1). The mean income of the local communities when equated with the mean operational holdings clearly indicates a direct relation with significant variations.

Table 1
Mean and standard Deviation of annual income and land holding (local community)

	Mean		*Standard Deviation*	
Local communities	*Annual income (in Rs)*	*Land holding (in ha.)*	*Annual income (in Rs)*	*Land holding (in ha.)*
Tribal	19018	0.1347	14591	0.2095
Non-tribal	30079	0.4180	27524	0.5709
Total	25167	0.2927	23349	0.4698

The human development indicators of the study area, that is, literacy rate, per capita income, health status, basic entitlements, sex ratio, work participation rate, accessibility to infrastructural facilities, and availability of social security options etc. give a comparatively low level of human resource development among the local community.

Poverty and Inequality

The incidence of poverty in the study areas was measured fixing the poverty line at the Monthly Per capita Consumption Expenditure (MPCE) of Rs. 335.46 (Anitha and Muraleedharan, 2002). The incidence of poverty is found to be high among the tribals while the non-tribals recorded a low rate (Table 2). It is also noted that the inequality among the tribals is much lower than that of non-tribals. This is due to the high level of disparities in the distribution of income and assets of non-tribals. There are large landowners and mere labourers among the non-tribals. In the case of tribals low inequality shows that there is no development to any segment of them i.e. all are facing with the same problem of lack of entitlements or assets.

Table 2
Poverty and inequality among the stakeholders

Group	*Poor (%)*	*Inequality index*
Tribes	56.89	0.0108
Non-tribes	42.14	0.2998
Total	49.52	0.3191

Source: Primary data estimates

Socioeconomic Dependency on Forest Resources

The local communities in the study area depend on the sanctuaries for their subsistence and complementary or supplementary uses. These dependencies may be in the form of collection of food, fuel, fodder, green manure, poles, NTFPs, fish, cultivation, water source, or catering to their social needs like cultural and religious. Based on these, dependencies can be classified as social and economic. The dependencies of the tribes and non-tribes vary considerably (Fig. 1). The tribal dependence varies from

13.2 per cent in manure to 77.4 per cent in fuel wood. In the case of non-tribals it varied from zero dependence in cultural, medicinal and food to 71.9 per cent in fuelwood. Tribals outweigh non-tribals in the case of overall forest dependency.

Figure 1
Forest dependence of local communities on Peechi-Vazhani and Chimmoni WLSs

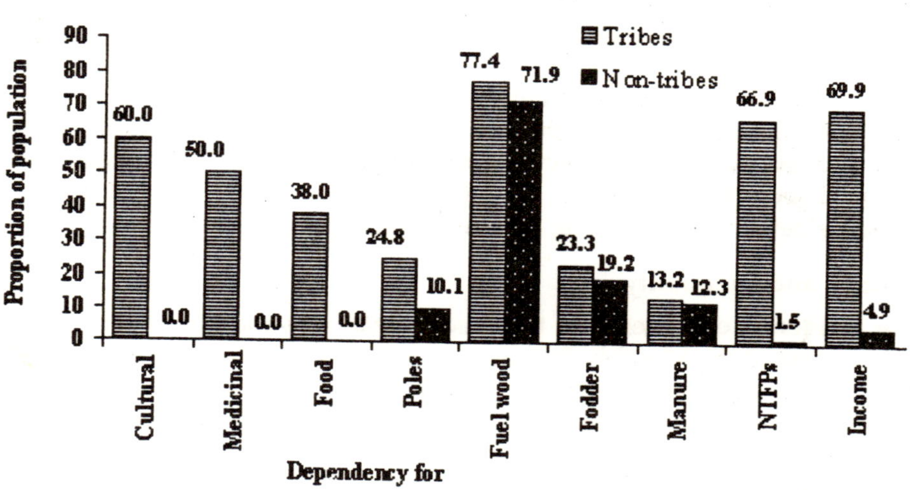

The Forest Dependency Index (FDI) was calculated to strengthen the analysis of forest dependence of the local communities on the WLSs. It has been found that the dependency level is significantly high among tribes (Table 3) both in the social and economic front indicating the significance of forestry in their livelihood.

Table 3
Forest Dependence Index of the local communities (Peechi-Vazhani and Chimmoni WLSs)

Dependence/ Community → *Parameter ↓*	*Dependency Ratio a (d/100)* *d = Dependency level*		
	Tribal(α_t)	*Non Tribal (α_n)*	*Total β_1(Average)*
Social dependence	0.55	0.00	0.28
Economic dependence	0.45	0.17	0.31
Total β_2 (Average)	0.5	0.09	0.30
FDI (β_2*100)	50.00	9.00	30.00
χ^2 value		**82.940**	

Primary data estimates
FDI: Forest Dependence Index.

Calculated value of the chi-square indicated the existence of a significant association between the social group and forest dependencies. It implies that dependence on forests of the individual depends on the social group they belong to. Forests thus provide a range of benefits in the form of goods and services that arise from direct and indirect use. Availability of various livelihood inputs (such as subsistence goods, income, employment, etc.) from the PAs helps in bridging seasonal gaps and determines the standard of living of the dependent communities particularly tribals. Ease of access and proximity to widely dispersed rural markets enable very large number of people to generate some income from forest produce.

IMPLICATIONS

The dependencies/pressures (Fig 2) on the WLSs have many implications. People living within and adjoining Peechi reservoir, mainly to supplement their family income, undertake cultivation in the catchment area, hence affecting river flows. The cultivation practice adopted by the local communities is highly unsustainable, particularly with the regular application of dangerous pesticides like *Phorate*, which is highly poisonous and known for its *'bio-magnifier effect'*. It is hazardous to humans, cattle and wildlife and environmentally threatens the sustainability of the environment, forest, wildlife, water and water catchment area as a whole (CPP, 1986). This is a very serious matter, needing immediate attention. The danger lies in the contamination of the reservoir when it fills with water immediately after the cultivation period, as it is the source of drinking water to Thrissur (the nearest town) and nearby panchayaths. In the Vazhani and Chimmoni catchment areas of the reservoirs no cultivation is observed.

There is a rich fish fauna in Peechi-Vazhani and Chimmoni WLSs, with a total of 37 species, belonging to 15 families (Thomas, et. al., 2000). A decline in the quantity of fish collected from Peechi reservoir is observed from 32,854 kg in 1991-92 to 4,224 kg. in 1998-99. Whereas, the quantity of fish collected from Vazhani reservoir has indicated an increasing trend over the years from 6,584 kg in 1992-93 to 13,060 kg in 2000-01. This may be attributed to the unsustainable land use practices undertaken in the Peechi reservoir catchment area and the less disturbed catchment area of Vazhani reservoir.

In the case of NTFP resources, important constraint faced from the conservation point of view is the destructive commercial extraction practices in response to market-driven demand. The imperfections in market structure and non-functioning status of the TSCS (in Peechi Vazhani WLS) makes the situation worst with enhanced and uncontrolled exploitation opportunities by private traders. In the study area 62 per cent of the NTFPs collected (classified based on use) is for roots, stem, bark

Figure 2
The Actual Commercial Threats

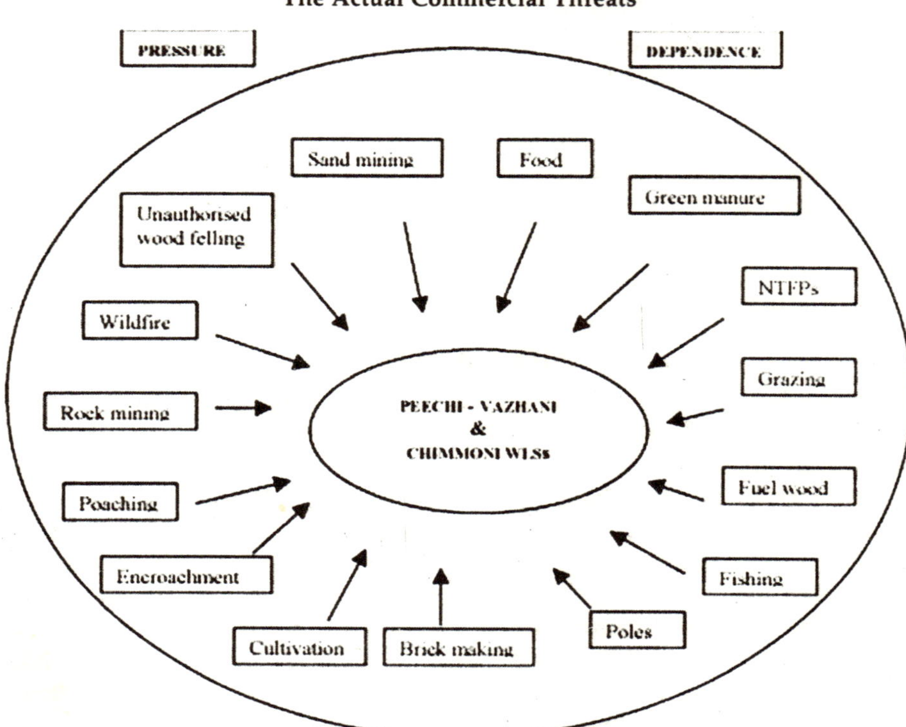

and whole plant thereby having potential threat of destruction to the NTFP species (Anitha and Muraleedharan, 2002). The commercial importance of the products can lead to degenerative extractive practices of the natural resource, if utmost care is not adhered to. A step-wise regression analysis is attempted in the study (Anitha and Muraleedharan, 2002) between income from NTFPs and 10 selected socio-economic variables (viz. the family size, age of the head of the family, income from wage labour, income from agriculture, asset value, distance traveled for collection, number of adult members, number of educated adults, education of the head of the family and experience in foraging). The only variable found to have a direct and positive influence on the income from NTFPs is the distance traveled; i.e., 59 per cent of the variation in the income from NTFPs seems to be controlled by the distance traveled (Table 4). When distance traveled is more, the proportion of NTFPs collected is also more. For instance, within the radius of less than 3 km, a collector is able to collect only 10 per cent of the total collection of a day and a collector has to travel at least about 10 km a day for one day's collections (Table 5). This further supports the argument that there is depletion of NTFP resources in the study area. More over the phenomenon of economic extinction, that is, the extinction of economic significance of the NTFPs from the collector's point of view due the high relative costs involved (visible and invisible) in the collection of the NTFPs and relative low benefits accrued to the collectors. About 48 per cent of the available commercially important NTFPs are facing the economic extinction in the study area. The economic extinction of non-premium NTFPs and the alternate sources of income have altered the level of dependency of forager households on NTFPs for their livelihood.

Table 4
Multiple Regression Analysis (Results)

Data group	*Results*
Group income from NTFPs Less than 50 per cent	$Y = -6359.3148 + 0.41464\,x_7 + 0.00744\,x_8 + 448.7026\,x_9$ (2049.13588) (0.16138) (0.00274) (174.19765) (Adjusted $R^2 = 0.4876$)
Group income from NTFPs , more than 50 per cent	$Y = -8732.44276 + 1731.78588\,x_9$ (1235.37145) (132.86479) (Adjusted $R^2 = 0.70111$)
Pooled data	$Y = -8174.12594 + 1736.83864\,x_9$ (1185.77893) (106.0338) (Adjusted $R^2 = 0.5949$)

Figures in parenthesis are standard errors
Y = HH income from NTFPs (Rs/Y); x_7 = Income from wage labour; x_8 = Total asset value
x_9 = Distance traveled for collection

Table 5
Per cent of NTFPs collected in different foraging distances

Distance covered	*Proportion of NTFP collection*
< 3 km	10%
3 km to 10 km	80%
> 10 km	100%

Source: primary data estimates

Forestland is also used for grazing mainly because of the scarcity of common grazing lands and the high cost of fodder. The grazing pressure in the study areas is recognized as approximately 25 livestock/km². The total grazing impact zone is 125 km² (Peechi – 75 km² and Chimmoni – 50km²). The grazing pressure in Peechi-Vazhani is significant and is closely correlated to forest degradation quality–wise (Jayanararyanan, 2000).

Fuelwood collection is also rampant, although it is a restricted activity in a PA to meet the subsistence needs, this is permitted to a large extent. It was estimated that about 74 per cent of the selected households collect firewood from the forests both for subsistence and sale. Fuelwood species generally collected are Chadachi (*Grewia tiliaefolia*), Irul (*Xylia xylocarpa*), Maruthu (*Terminalia paniculata*), Thek (*Tectona grandis*), Venthek (*Lagerstroemia microcarpa*) due to high demand in the market. It was estimated that about 1,400 tonnes of firewood is collected annually, of which 560 tonnes are for sale and 840 tonnes for self-consumption. The sale price of one bundle of firewood (20 to 35 kg) at the market varies between Rs. 35/- and 70/-, depending upon area, season, demand, etc. Unrestricted extraction of fuelwood particularly for commercial purpose has become a threat to conservation of tree species.

Other commercial interests like sand mining and brick making are also being pursued here in the Peechi catchment area. Each year, for four months (December-March) six brick making units function in the Peechi catchment area. Each unit produces about 15 lakh bricks and hence makes an annual turnover of about Rs.16 lakhs. About 60 families living near the area are completely dependent on these brick making units for their livelihood. Rock mining is another activity being undertaken in the PA exercising immense pressure on the PA and indirectly polluting (noise as well as material waste) the atmosphere. This is attributable to the socioeconomic conditions of the local communities, lack of alternative job opportunities, negligence on the part of government officials and the political support to labour unions.

Unauthorised activities such as encroachment especially for the purpose of cultivation also take toll on the PAs. The issue of encroachment is highly complex and often seemingly irresolvable. Approximately less than 180 hectares of land is in the possession of the local community within the Peechi sanctuary with around 15% alone having proper title deeds (Jayanarayanan, 2000). Poaching, although an unauthorised activity is found only in certain areas (eg. Karadippara, Chakkolatharissu, Pullankandam, etc.), again for self-consumption and often traded. This contributes to the degradation process.

These adverse anthropogenic pressures along with various unauthorised activities are the actual commercial threats that these sanctuaries face today wanting immediate action.

CONFLICTS IN PA MANAGEMENT – THE CAUSATIVES

All forestry interventions deal with a historical interface between people and the landscape, their mutual influences, which is often considered less significant in any management plan. Of particular impact are the links between forestry and the basic needs of forest-dependent communities. There exists a diverse conflict in the study areas. They are basically (i) *for land* (change in boundaries, wanting tenurial security); (ii) *for produce/ resource*, either for self-consumption or for sale (either legitimate – eg. cheenikka, or illegitimate – eg. rose wood, deer meat); and (iii) *for political dominance*, exhibiting extremist behaviour/attitude. These are found on outer boundaries (eg.Thirumani, Kalappara, etc.), along inner boundaries (eg. Maniyankinar, Thamaravellachal, Olakara, etc.) and between social groups (tribes, *malayans, kaders*; non-tribes *kavaras, pulyas, vettuvas* and *parayas;* and others; also between Forest Department and local communities). Existence of human enclaves within the WLS (eg. Maniyankinar, Thamaravellachal, Mundipadam) is a potential threat in the long run as they tend to expand at the cost of the sanctuary. There are a wide range of conflicts and disputes adversely influencing PA management of the study areas.

There are a number of factors that have contributed to the natural resources based conflicts in Peechi-Vazhani sanctuary. Some of the causes are the attempts to evict the encroachers/forest dwellers, changing socio-economic needs, lack of community awareness, traditional institutional arrangements that governed resource conservation and use, political interference and the growing population and dwindling resources. Some of the causative factors of human-related constraints in the study area, most of which are more prominent in Peechi-Vazhani than in Chimmoni, are discussed here.

Property rights are not well defined. Thus, conflicts arise over land ownership, allocation of land, lack of access roads, and sometimes the land use practices. Undefined tenure issues lead to exploitation of natural capital (*open access resources*). Most of the people living next to or inside the forests do not know the provision of the laws as to their rights and responsibilities to forest conservation. They, therefore, strive to use the forest without taking sustainability into account.

The increase in the population in the surrounding areas of the PAs increased the pressures on the PAs, which lead increased extraction of its dwindling resources. The latest decadal population growth rate of over 8.7 per cent means that pressure on PAs will continue. This will lead to a progressive reduction of forest cover. Unclear institutional arrangements end up confusing the resource-dependent communities. There is a lack of coordination among the various institutions like Forest Department, Water Authority, Irrigation Department, Tribal Service Co-operative Society, Public Works Department, etc which have a stake on the study area in question.

The policy and legal framework are not in line with the changing demands of local communities. This results in uncontrolled exploitation of the resource by forest adjacent communities whose livelihood depends on the forests. Forestland is being allotted to people who end up putting it to different uses that may have a negative impact on environment. Political intervention also led to irregular forestland allocation with sustainability of the resulting settlements not guaranteed. As a result of the 'globalisation' process the local resources are exposed to the outside market due to internal and external demands. This implies competition over limited resources. Local communities who cannot afford to compete in a liberalized market, yet depend on natural resources for basic needs, have no other option but to over-exploit resources.

PROTECTED AREA MANAGEMENT – FEASIBLE OPTIONS

There is an urgent need for decision-makers to participate in debates and contribute to issues on how to counter conflicts by doing away with issues pertaining to natural resource management. Given that a conflict cannot operate outside social context, it is necessary to adopt a management strategy where along with the PA objectives, sustainable livelihood issues are also addressed or human dimension should be given the due importance. In other words, it is essential to adopt an integrated landscape-livelihood approach to PA management. In a landscape there are several elements like the forests, degraded forest, village ecosystems, etc., which have inter-linkages. The changes that happen now and potential changes can be understood and managed only when the human interactions are studied and understood. However, changes due to management decisions and human interactions are quicker and there is a need to manage such changes in a way that does not compromise the long-term sustainability of the resource base and the options available. A sustainable landscape-livelihood approach (Anitha and Muraleedharan, 2002) is the need of the hour in removing constraints and exploiting opportunities to realize positive livelihood outcomes at the community level.

CONCLUSION

The multiplicity of forest products and their uses, the conflicts they may cause and the various other commercial interests make forestry planning at the sector level highly complex. The current scenario existing in the study areas highlights a situation where the local communities have little say over the management and policy matters pertaining to conservation. Resource constraints and poor socioeconomic conditions coupled with easy accessibility to the WLSs explain the anthropogenic dependence/pressure on the study areas. This is further complicated with the skewed distribution of resource endowments. The high incidence of poverty among tribals coupled with their relative high degree of dependence on the WLSs is indicative of the threat of further degradation. Human consumptive use of resources and human land use practices are incompatible with conserving biological diversity. The existence of enclave settlements, which tend to expand at the cost of the sanctuaries, is a potential threat in the long run. The decreasing income potential from forest-based activities has forced them to look upon non-forest based activities, especially cultivation, which has led to land-based conflicts. The tenurial complexities in the sanctuaries need to be addressed in order to resolve the land-based conflicts. As the dependence on NTFPs by the local community particularly tribals as one of the main income source is very high, sustainable management of NTFPs in the study areas should be achieved through a combined attempt of strengthening appropriate institutions with adequate management backup, participatory management involving the actual foragers, judicious use of products, adoption of scientific methods of extraction and value addition and practices of *in situ* and *ex situ* conservation. It is suggested to adopt landscape livelihood approach as the PA management strategy where along with the PA objectives sustainable livelihood issues or human dimensions are also addressed in a participatory mode of management.

REFERENCES

Anderson, A. (ed.) (1990), *Alternatives to Deforestation: Steps Toward Sustainable Use of the Amazon Rain Forest*, Columbia University Press, New York.

Anderson, John, Gauthier, Michelle, Thomas, Garry and Wondolleck, Julia (1996), Addressing Natural Resources Conflicts Through Community Forestry: Setting the Stage, Paper Prepared for the Electronic Conference, Forest Trees and People Programme, Forestry Department, FAO. Rome, Italy, p. 10.

Anitha, V. and Muraleedharan, P. K. (2002), Study of Social and Economic Dependencies of Local Communities on Protected Areas: A Case of Peechi-Vazhani and Chimmoni Wildlife Sanctuaries, *KFRI Research Report, 240,* Kerala Forest Research Institute, Peechi, Thrissur, Kerala, India.

Arild, A. (2001), Playing Games in the Forests: State Local Conflicts of Land Appropriation, *Land Economics*, 77(2): 285-299.

Chambers, R. (1997), *Whose Reality Counts,* Intermediate Technology Publications, London.

Claude, D. and Michelle, G. (2000), Community Forestry and Forest Resource Conflicts: An Overview, Internet File, http://www.fao.org/forestry/foda/wforcong/PUBLI/PDF/V5E_T27.PDF

Colfer, C. J. Pier (1995), Who Costs Most in Sustainable Forest Management, Working Paper No. 7, CIFOR, p. 15.

CPP (Chemical and Pharmaceutical Press) (1986), *Crop Protection Chemicals,* Reference 2nd Edition, Chemical and Pharmaceutical Press, John Wiley and Sons, New York, p. 102.

Gadgil, M. (1992), Conserving Biodiversity as If People Matter: A Case Study from India, *Ambio*, Vol. 21, No. 3: 266-270.

Gutman, P. (2001), Forest Conservation and the Rural Poor: A Call to Broaden the Conservation Agenda, WWF Macroeconomics for Sustainable Development Programme Office (MPO), Washington DC.

Hatch, C. (1997), *Consuming Canada's Forests, Brazil of the North II,* p. 23.

Jayanarayanan, T. (2000), Forest Degradation in Kerala - Causes and Consequences: A Case Study of Peechi-Vazhani Area, *Discussion Paper No. 27,* Kerala Research Programme on Local Level Development (KRPLLD), Centre for Development Studies, Thiruvananthapuram.

Kutty, R. and Kothari, A. (2001), Protected Areas in India – A profile, Kalpavriksh, Pune, India.

Mammen, and Deepa. P. (2000), Landscape Evaluation of Peechi-Vazhani Wildlife Sanctuary with Special Reference to Peechi Region Using Remote Sensing Techniques, MSc. Dissertation (unpublished), School of Environmental Studies, Mahatma Gandhi University, Kottayam, Kerala, India.

McCleery, D. (2000), Searching for the Underlying Causes of Forest Loss and Degradation: Dealing with Issues of Scale and Context, Professional Paper (Internet) <http://www.bionet-us.org/docs/uc-na_maccleery.pdf>

Michael W. (2000), Conflict Management in Community Based Natural Resource Projects: Experience from Lakekamu Basin Integrated Conservation and Development Project, Papua, New Guinea, *Biodiversity and Ecological Economics – Participation, Values and Resources Management*, Earthscan Publications Ltd., London and Sterling, VA.

Narayanankutty, T. P. (1990), The First Management Plan for Peechi-Vazhani Wildlife Sanctuary 1990-99 to 1999-2000, Government of Kerala, Kerala, India.

Palik, S. (1993), Joint Forest Management, *Working Paper No. 15*, National Support Group for Joint Forest Management and the Ford Foundation, New Delhi.

Poffenberger, M. (1990), Forest Management Partnerships – Regenerating India's Forests, The Ford Foundation, New Delhi.

Thomas, K. R., Biju, C. R. and Ajithkumar, C. R. (2000), Fishes of Chimmoni and Peechi-Vazhani Wildlife Sanctuaries, Kerala, *Indian Journal of Bombay Natural History Society*, Vol. 97, No. 2: 289-292.

34

THE PUZZLE OF IDLE LAND IN THE DENSELY POPULATED KIGEZI HIGHLANDS OF SOUTWESTERN UGANDA

Bamwerinde W., B. Bashaasha, W. Ssembajjwe and F. Place

INTRODUCTION

The development of the agricultural sector in Kigezi highlands in South Western (SW) Uganda has involved progressively more intensive use of land resources for cropping and grazing leading to unsustainable adaptation and changes in land use. Restoration of soil productivity in Kigezi highlands had relied on natural fallow practices and subsequent natural resource regeneration. However, existence of large proportions of uncultivated land in such densely populated area is evidence that some of this may be unintentional fallows (AFRENA, 1999). What is observed appears contrary to the expectation of increased annual cultivation in land scarce areas as postulated by the theory of induced innovation (Boserup, 1965). Some authors have suggested that the increased area under long fallows could be a farmer's management strategy to improve natural resource management and particularly fertility regeneration (Lindblade *et al.*, 1998). Others have argued that disuse is because plots are very heterogeneous with some being very unproductive in terms of crop yields despite effort to increase production. Another explanation is that poor land markets could be inhibiting farmers from acquiring land that they would put into use and disposing of land that they have abandoned. Despite the above explanations, the economic determinants, causes and extent of land use decisions by farmers are not well understood.

As a result, there are as yet no agreed upon interventions to address the issue. In order to understand the impact of human activity on the highland ecosystem, it is important to identify and estimate the factors that influence farmers' decisions to put land to particular usage. Therefore, the main objective of this study is to determine the present status and causes of land abandonment in Kabale District, and propose policy options for sustainable management of land resources.

The Kigezi highlands are a region of high agricultural potential but with a high population density (about 300 per km^2) and population growth of about 2.2% per annum (Ministry of Finance, Planning and Economic Development, (MFPED, 2004). The highlands lie in SW Uganda at an altitude of 1500-2700 meters above sea level. The rainfall pattern is bimodal and ranges from 1000 to 1500 mm per annum. The temperatures are moderate with mean minimum of 13° C and mean maximum of 23° C. In Kigezi, soils are generally volcanic and were inherently fertile although some parts have less fertile Ferrasols and Andosols (Djimbe and Hoekstra, 1987).

PREVIOUS WORK ON FARMERS LAND USE DECISIONS

In this research, the practice of leaving land to rest for longer periods than the farmer would have desired (*unintentional fallow*) locally known as "*Eibija*" is what is referred to as abandonment of land. This is opposed to *intentional fallow* that is commonly practiced and known to farmers locally known as "*Hinga raza.*" Land abandonment has been observed in several other countries including Botswana, Japan and Lebanon (Mmopelwa ,1998; Arntzen, 1984; Kashiwagi, 1998; Zurayk 1994).

Mmopelwa (1998) empirically examined the proportion and factors causing unintentional fallowing in South East Botswana. He used data from 100 randomly selected farmers and simple statistical estimations. The study revealed that 20% of farmers unintentionally fallowed (short-term or long-term), 1% intentionally fallowed and 79% did not fallow at all. The factors attributed to unintentional fallows were the lack of draught power (86%), labor (82%) and rainfall variability (100%). Based on these findings two aspects of fallowing were noted. The first aspect was intentional fallowing, in which farmers purposefully left their land uncultivated for it to regain soil fertility. The second aspect was that of unintentional fallowing whereby farmers involuntarily leave their land uncultivated due to biophysical, social and economic factors beyond their control.

Grisley and Mwesigwa (1995) investigated the socio economic factors that influence seasonal fallowing in Kigezi highlands. The study revealed that 76% of farmers reported some cropland in grass fallow. Households reported an average of 26% of cropland under fallow. Model estimates revealed that intercropping and farm size were highly associated with the land fallowing decision but distance to fields was not. They recommended use of capital-intensive technologies such as terracing, agroforestry and use of chemical fertilizer, though currently unavailable, to overcome the problem of land being idle for a long time.

An empirical study by Barbier (1998) used a bio-economic model to investigate the theory of induced innovation and land degradation in the intensively cultivated areas of West Africa. The results of this study show that farmers prefer extensification to intensification even if land is marginal and produces lower yields. Under population pressure, the model revealed that over time soil organic matter and soil nutrients become scarce and the cost of land degradation increases. However, if the population is constrained and continuously farms the same plots, the economic results and welfare deteriorates especially when existing markets and technologies for agricultural improvements are limited.

The changes in land use system in Kabale District in Kigezi highlands between 1945 and 1996 as well as the current farm management practices were investigated by Lindblade *et al.* (1998) using transects conducted in 1945 as baseline for measuring land use changes. They observed significant increase in area under fallow since 1945. For example, 19% of plots were left to fallow in 1945 as compared to 32% in 1996. Furthermore, the length of the fields rested had extended significantly with 95% left to fallow for more than half a year in 1996 compared to only 5% in 1945. Households with more than 11 plots reported 31% of their plots under fallow while farmers with less than 5 plots could fallow up to 25%. Finding more fallow to day compared to 1945 is perplexing given the perceptions of the local people. Rapid appraisal exercises conducted by various researchers reveal consistently that the perception of local people is that land fallow periods have decreased or are non-existent for some farmers. One reason could be due to definitions and assumptions: Lindblade *et al.* (1998) considered all uncultivated plots as fallowed.

ANALYTICAL FRAMEWORK

The five land use options analyzed in this study include continuous cropping, short fallow, long fallow, woodlots and abandonment. At an individual farm level, decisions on whether to use a given plot of land for annual production or to put it under other land use options such as long fallow and woodlot is assumed to depend on the net present value of the option available to the farmer. For example, the practice of continuous cultivation with intermittent short duration fallows would only continue if normal returns on fixed capital would be obtained after all farm running costs have been paid. Continuous crop production and short fallow involves a modest capital investment, are short term in nature (0-1 year), and can result in optimal use of the plot if well managed (e.g. with adequate nutrient inputs). However, if land resources become depleted, potential future profits will be reduced until ultimately the plot is fallowed or put to some other land use option that would be less profitable in comparison to cropping if land was not degraded. Because the use of fertilizer is very low in Uganda as a whole, Kigezi no exception, the possibility that land become degraded with continuous use is real.

Fallowing is a land management strategy to intentionally increase soil productivity in the medium term (1-3 years). Consequently, land fallowing is an investment decision. This study argues and tests the proposition that not all farmers can afford this medium-term investment and therefore some of the plots that appear to be in fallow are in reality abandoned because of unintentional sub-optimal usage by farmers. The complexity between plot abandonment and long fallowing is that both practices appear to be indistinguishable by observation but based on different decisions. Therefore, long fallow, woodlots and abandonment decision by farmers are treated as different forms of long-term investment.

We assume that plot characteristics that affect productivity or the net returns will affect land use decisions. For example, plots of inherently poor soil fertility would be less likely to be put to into cultivation compared to other plots. The distance to the plot will also generate differences in labour costs and risks (e.g. of crop theft) and affect land use decisions. Household factors such as access to resource (e.g. labour, land) will also affect the net returns from the portfolio of plots and affect general land use decision-making at the farm level. As will be shown below, farms are characterized by extreme land fragmentation, with households owning 6-10 plots widely scattered on the landscape. Thus, the possibility that a household will face widely varying net returns to an identical investment on their different plots is high. Farmers may also opt to abandon the plot after a short fallow. In some cases, farmers could revert land previously abandoned back to cropping depending on the availability of inputs.

ECONOMETRIC MODEL OF LAND USE DECISIONS

We estimate an econometric model of land use decisions using a snapshot or cross sectional data. The dependent variable (land use options) has five response levels representing the current land use option of a given plot. These are continuous cropping, abandonment, woodlot, long fallow and short fallow. The covariates include age of the household head, farm size, plot distance from homestead, household size, education level of the household head, gender of household head, polygamous or not, plot location, soil fertility status, presence of stones, plot size and slope. The logistic model is generally specified as follows;

$$Y_i^* = \beta' X_i + \varepsilon_i \tag{1}$$

Where, Y_i^* is the underlying latent variable that indexes the land use option on a given plot. X_i is a $(k \times 1)$ vector of explanatory variables, β is a $(k \times 1)$ vector of the parameters to be estimated, and ε_i is the stochastic error term.

A general expression (Greene, 2003) for the conditional probability for the five land use options (categories) is:

$$P(Y = j) \Big| X = \frac{e^{g_j(X)}}{\sum_{k=0}^{4} e^{g_k(X)}} \tag{2}$$

Where j is the category of land use. From this specification, the probabilities for each of the observed responses are easily derived.

Empirical Model

The model used to estimate the influence of plot and household characteristics on land use choices is specified as follows:

$A = \beta_0 + \beta_1$ FARMSIZE $+ \beta_2$ HHSIZE $+ \beta_3$ AGEHH $+ \beta_4$ HHPMALE $+ \beta_5$ HHFEMALE $+ \beta_6$ EDUCPRIM $+ \beta_7$ EDUCPOST $+ \beta_8$ FERTMOD $+ \beta_9$ FERTGOOD $+ \beta_{10}$ TOPMID $+ \beta_{11}$ TOPTOP $+ \beta_{12}$ STONFEW $+ \beta_{13}$ STONMANY $+ \beta_{14}$ SLOPMOD $+ \beta_{15}$ SLOPSTP $+ \beta_{16}$ PLOTSIZE $+ \beta_{17}$ PLODIS $+ e$

Where;

A = Land use option (continuously cropped plots, short fallow, long fallow, abandonment and woodlots).

FARMSIZE	The total land holdings measured in hectares
HHSIZE	Number of household members measured in number of persons
AGEHH	Age of the farmer measured in years
HHFEMALE	Households headed by female (base case is monogamous households headed by males)
HHPMALE	Polygamous households measured as number of women living with the head.
EDUCPRIM	Household heads with education level of primary (base case is no formal education)
EDUCPOST	Household heads with education levels of post-primary or more
FERTGOOD	Terrace plots with good soil fertility
FERTMOD	Terrace plots with moderate soil fertility (base case is plots with poor soil fertility)
TOPMID	Terrace or plot is located in the mid slopes
TOPTOP	Terrace plots located on landscape (base case is valley bottoms)
STONFEW	Plot has few stones (base case is plots with no stones)
STONMANY	Plot or terraces has many stones
SLOPMOD	Plot or terrace is on moderate slope (base case is flat plots)
SLOPSTP	Plot or terrace is on steep slope
PLOTSIZE	The plot size in hectares
PLODIS	The distance of the plot from the home measured in Km

Selected Variables for Model

Several exogenous or predetermined household variables are included in the model. The age of the household head is included because the probability of plot abandonment may differ with age. However, the relationship may move in different directions. Land markets exist, but are imperfect so older farmers may have had more opportunity to dispose of unwanted plots or have accumulated more wealth so as to avoid abandonment. On the other hand, they may have acquired many plots and found that they cannot adjust their land holdings as quickly as other factors like labor or capital have changed.

Household size, sex of household head, polygamous households, and education of household head are also included. Household size is hypothesized to have a negative relationship with abandonment. Female-headed households are hypothesized to be more prone to abandonment than male-headed households because of their difficulty in exercising rights or acquiring necessary factors of production. Educated farmers (especially those with post-primary education) are expected to have acquired knowledge in relation to improved methods of land management. They also have the ability to acquire loans to improve the land. However, they are also capable of diverting the resources to non-agricultural activities perceived to offer higher returns. This therefore may lead to less investment in soil improvement and increased land abandonment. Therefore, a priori the hypothesized sign is unclear.

The plot related variables included in the model were TOPTOP (plot on a hilltop) and TOPMID (plot on a mid-slope) with plots in the valley bottom acting as a base. Although both represented the topographic location of the plot, they are entered as separate variables because they are expected to influence abandonment of plots differently. It is expected that plots located on the hilltop are more likely to be abandoned compared to plots in the valley bottom because of poor accessibility. The plots in the mid slope are more likely to be either short fallowed or long fallowed because they are closer to the homes. The degree of soil fertility of a plot influences land use decisions. The variables FERTMOD (moderate soil fertility) and FERTGOOD (good soil fertility) are entered separately as dummy variables and plots with poor soil fertility acted as the base case. It is hypothesized that soil fertility is negatively related to abandonment. Therefore, both soil fertility variables are both expected to have negative signs.

The influence of plot stoniness on farmers' decision to abandon or cultivate the plots is also investigated. STONFEW (few stones) and STONMANY (many stones) are variables entered separately because many stones are hypothesized to encourage abandonment since it is difficult to open the soil (absence of stones is the base and omitted case). Absence of stones is expected to encourage continuous use of the plot. Therefore, the hypothesized sign is positive for STONMANY and STONFEW, but weaker in the latter case. The slope of the plot is also a characteristic expected to influence the use of the given plot. The slope variables are SLOPMOD (moderate slope) and SLOPSTP (steep slope). They are entered as separate variables and flat plots are used as base cases to verify the influence of slope on abandonment. The steeper the plot, the more difficult it is to cultivate and improve, hence greater slopes are expected to have a positive sign.

The hypothesized sign is positive for FARMSIZE and negative for PLOTSIZE. The larger the farm, the less likely it will be for a farmer to keep all of it in continuous cultivation, ceteris paribus. Larger plots have a negative sign and are expected to act as incentives for farmers to travel long distance and use the plots despite fragmentation. The variable (PLOTDIS) represents the distance from the homestead to the plot. It is hypothesized to have a positive sign. The further away the plot was from the home, the higher the probability of it being abandoned or long fallowed.

Data and Sources

To estimate the model, data were collected from 3 sub counties in Kabale district. Kabale district with a total land area of about 1850 km^2 has three counties and 19 sub counties. In each county, two sub counties were selected making a total of six sub counties. These included Kashambya, Kamwezi, Rubaya, Maziba, Bubale and Hamurwa sub counties. The unit of analysis in this study was a farmer's plot, but data were also collected on the farming household. Since factors related to plot abandonment were analyzed, a plot was used because it is the smallest unit of land on which cultivation takes place.

In order to allow for a wider coverage and representativeness, a random sample of respondents was used. Selection of sub-counties and parishes was undertaken using multistage simple random sampling. Twenty farmers were randomly chosen from each locality while 60 farmers were taken from each sub county. Interviews were carried out in 18 parishes. Data were collected from a total sample of about 360 households through use of a structured questionnaire and plot visits. Given the number of plots per household in the district, over 2,530 plots were covered in the survey.

EMPIRICAL RESULTS

Extent of Abandonment and Land Use Options

The current use of plots by farmers reveals that about 59.6% of the 2,530 plots are cultivated with crops. The short fallow and long fallow comprise of 8.1% and 13.7% respectively. The abandoned plots by farmers are about 10.1% of the total plots visited. The plots with woodlots comprise about 8.1%.

Farmers' reported that poor yields are the main reason for both long fallow and abandonment of fields. The survey reveals that of roughly one-quarter of plots either long fallowed or abandoned, 61% was due to soil fertility problems. Livestock grazing (10%), location of fields (11%), labor shortage by households (9%) and old age (2.3%) are also important factors for non-cultivation. Other reasons given include bush fires and farmers poor health that affected about 3% of the plots surveyed.

The results (Table 1) reveal that the average distance of plots from the homestead and roads is about 1.15 km and 1.5 km respectively ranging from 0 to 20 km. Because of the terrain, even when distances are small, it is difficult to reach the plots that look to be near homes. The mean distances to market from plots was 6 km. The average number of plots owned by the household head is 8 plots and ranges from 1 to 20 although scattered over the landscape. This scattering of land though sometimes used as a risk management strategy may partly lead to sub optimal usage of the plots.

Table 1
Variations in Factors that Influence Farmers' Land Use Decisions

Factors	*N*	*Minimum*	*Maximum*	*Mean*	*± S.E*
Distance of plot from home	2530	0.00	20.00	01.15	0.03
Total number of plots	2530	1.00	20.00	07.80	0.08
Farm size	2530	0.07	46.00	10.97	0.25
Age of household head	0360	18.0	82.00	45.12	0.27
Household size	0360	1.00	23.00	07.73	0.00
Size of plot	2530	0.03	25.00	01.26	0.04
Area under woodlot	2530	0.00	27.00	00.98	0.148
Area under short fallow	2530	0.00	25.00	00.68	0.119
Area under fallow	2530	0.00	26.00	01.23	0.149
Area abandoned	2530	0.00	57.00	01.07	0.193
Area under crops	2530	0.00	60.00	04.88	0.014

Note: Distance was measured in km, farm size and area in acres.

The mean household farm size is about 10 acres (4ha). This appears to be on the high side compared with existing data, which reports the average land holding as less than 1 hectare. Examination of data and the land use system suggest that an average land holding of less than 1 hectare per farmer in the Kabale district could be an underestimate. This appears to be based on the total land area divided by the total number of farm households, which does not cater for the land farmers own on the hillsides. Furthermore, unless one visits the plots, farmers tend to report fewer plots.

Farmers tend to report only plots they use gainfully rather than plots lying idle.

The mean for household size is about 7 members per household with only 4 active members (55%) who are above 14 years. The rest are either elderly (0.5) or still young (3) members. Apparently, although this could be a good source of farm labor (especially the young) given the fact that universal primary education is in place, this figure leaves few members available for full time labor.

Estimates of Determinants of Farmers' Decisions

A single multinomial logit model was run, but because there are five outcomes, the estimates of the determinants of farmers' decisions for each separate category are given in Tables 2 and 3. The coefficients can be interpreted as the change in the log odds associated with a unit change in the respective independent variable. They represent the degree to which a one-unit change in an independent variable induces a change in likelihood from a cultivated plot to one of the other four land uses. Results of the four land uses are presented in the following order, short fallow model, long fallow, woodlot and abandonment models.

Short Fallow Model

The household level variables that are significant for short fallow were FARMSIZE and EDUCPOST. Other significant plot variables included, SLOPMID, FERTMOD, FERTGOOD, PLOTSIZE and PLOTDIS. The short fallow model results show that the odds ratio values for a given plot to be under short fallow are significant for EDUCPOST and FARMSIZE at a 1% level. For example, the odds ratio values obtained for EDUCPOST is 2.26. This implies that as the variable changes from no formal education to EDUCPOST, the odds ratio for the plot to change from cultivated plot to be under short fallow increases more than 2 times compared to that of no formal education, the value of other independent variables held constant.

Table 2 Estimated coefficients and odds ratio for factors influencing farmers' decisions to either use land for short fallow or long fallow. Analysis was done using multinomial logit function against abandonment, woodlot, short fallow and long fallow based on continuous cultivation.

The plot variables such as TOPMID, PLOTDIST, and FERTGOOD significantly influence short fallowing. For example, the results show that the odds ratio of TOPMID was 2.06 and had a negative sign. This shows that the probability of a plot being under short fallow are 2 times less likely if a plot is located in the mid-slopes relative to the plots in the valley bottom. Other variables such as PLOTSIZE and FERTMOD are significant at the 5% level and have the hypothesized signs. The negative sign on plot size suggests that the larger the plots, the less likely they are put under short fallowed.

Table 2
Estimated Coefficients and Odds Ratio for Factors Influencing Farmers' Decisions (Land Use for Short Fallow and Long Fallow)

	Short fallow		*Long fallow*	
Factor influencing farmers	*Coefficient estimate of the HH/plot factor*	*Odds ratio*	*Coefficient estimate of the HH/plot factor*	*Odds ratio*
Intercept	–1.799		–0.767	
AGEHH	0.003	1.002	0.013	1.013**
HHSIZE	0.003	1.003	–0.059	1.060
FARMSIZE	0.026	1.026***	0.005	1.006
EDUCPRIM	0.118	1.125	0.217	1.240
EDUCPOST	0.816	2.260***	–0.013	1.014
HHFEMALE	0.105	1.110	0.488	1.630**
HHPMALE	0.026	1.027	0.445	1.550**
TOPTOP	0.299	1.348	0.279	1.321
TOPMID	–0.723	2.060***	–0.526	1.692***
FERTMOD	–0.486	1.625**	–1.657	5.243***
FERTGOOD	–0.877	2.403***	–2.276	9.738***
STONFEW	–0.230	1.258	–0.172	1.188
STONMANY	–0.359	1.431	0.031	1.031
PLOTSIZE	–0.133	1.142**	–0.017	1.017
SLOPMOD	0.086	1.090	0.225	1.250
SLOPSTP	0.297	1.346	0.294	1.341***
PLOTDIS	0.126	1.134***	0.195	1.215

*, **, *** denotes statistical significance at the 10%, 5% and 1% level respectively
HH denotes household

Long Fallow Model

Three variables related to household characteristics are significant at the 5% level. These include the age of the household head (AGE), households headed by females (HHFEMALE) and polygamous families (HHPMALE). Relationships between long fallow and age of household head have a positive sign. This implies that long fallow plots are more likely to belong to farmers who are older compared to those who are younger. The odds ratio value obtained for the variable (HHFEMALE) is 1.6. This shows that the probability (odd values) for a given plot to change from cultivated to long fallow is relatively higher for households headed by females compared to those headed by males. Furthermore, results show that the chance of long fallowing is greater for households with more than one resident wife compared to those with one resident wife.

The most significant variables are related to the plot. The significant differences are observed in soil fertility, location of the plot in the mid-slopes and plot distances at a 1% level of significance. For example, results show that the odds ratio for FERTMOD and FERTGOOD were 5.24 and 9.74 respectively and had the hypothesized negative sign. This implies that the odds ratio value of long fallowing for poor soils is 5 times higher compared to that for plots with moderate soil fertility and about 10 times greater than plots with good soil fertility. Again, the plots under long fallows are more likely to be located far from the homes compared to cultivated plots.

Woodlot Model

The woodlot model results (Table 3) reveal that one household variable (EDUCPOST) is important in determining the farmers' decision to establish woodlots and had the expected positive sign. Farmers with post primary education are more diversified into profitable livelihoods and could afford medium and long-term investments.

Table 3
Estimated Coefficients and Odds Ratio for Factors Influencing Farmers' Decisions (Land Use for Woodlot and Abandoned Plot)

	Woodlot		*Abandoned plot*	
Factor influencing farmers	*Coefficient estimate of the HH/plot factor*	*Odds ratio*	*Coefficient estimate of the HH/plot factor*	*Odds ratio*
AGEHH	0.011	1.012 *	0.020	1.020 ***
HHSIZE	0.019	1.020	0.219	1.245
FARMSIZE	0.097	1.000	–0.005	1.007
EDUCPRIM	0.414	1.513 *	0.781	2.184 ***
EDUCPOST	0.671	1.956 **	–0.058	1.060 **
HHFEMALE	–0.132	1.141	0.313	1.368
HHPMALE	–0.555	1.742 *	–0.398	1.489
TOPTOP	–0.118	1.125	1.073	2.924 ***
TOPMID	0.243	1.275	0.255	1.290
FERTMOD	–1.950	7.029 ***	–2.279	9.767 ***
FERTGOOD	–2.892	18.029 ***	–2.669	14.425 ***
STONFEW	0.184	1.202	–0.316	1.372 *
STONMANY	1.216	3.373 ***	0.419	1.520 **
PLOTSIZE	0.083	1.087 *	0.070	1.073
SLOPMOD	-0.030	1.031	0.739	2.094 ***
SLOPSTP	0.768	2.155 ***	0.522	1.685 **
PLOTDIS	0.094	1.099 *	0.211	1.235 ***

*, **, *** denotes statistical significance at the 10%, 5% and 1% level respectively.
HH denotes household

The plot variables that are significant at 1% level include, STONEMANY, FERTMOD, FERTGOOD, and SLOPESTP. Size of the plot is also significant although only at 10% level of significance. The fact that variables FERTMOD and FERTGOOD, reflecting plots with either moderate or good soil fertility are strongly negative and significant implies that soil fertility was one of the critical determinants for farmers' decision making to use plots for woodlots. Woodlots are more likely to be planted on plots with poor soils compared to fertile soils.

The stoniness is another plot variable that is highly significant. The odds ratio for stoniness in the woodlot model is 3.37. These results show that the odds for establishing woodlot increased to about 3.5 times among plots with many stones than the odds among plots with no stones. The steep slope variable is highly significant at 1% level. Farmers established woodlots on steeper plots as compared to gentler plots. The plot distance variable (PLOTDIS) represents the plot distance from homes. It had a positive sign but is significant only at the 8% level. The further the distance to the plots the more likely one observes a woodlot rather than a cultivated field.

Table 3 estimated coefficients and odds ratio for factors influencing farmers' decisions to either use land for Woodlot or abandon it. Analysis was done using multinomial logit function against abandonment, woodlot, short fallow and long fallow based on continuous cultivation.

Abandoned Plot Model

The significant variables explaining abandonment include both household and plot level variables. Table 3 highlights two household variables that are significant at 1% level. These are age of the household head (AGE) and primary level education (EDUCPRIM). Post primary (EDUCPOST) is significant at the 5% level. The odds ratio value obtained for EDUCPRIM is 2.18. This implies that as the variable changes from no formal education to EDUCPRIM while the value of other independent variables are constant, the (odds) relative chances of a plot being abandoned increases by about 2 times. The sign reverses for post-primary education. Farmers with abandoned plots are more likely to have post primary education compared to those with cultivated plots. Farmers with post primary education may have greater financial resources with which to invest on their land, as shown in the woodlot results.

The plot level variables related to abandonment that are highly significant at 1% level include TOPTOP, FERTMOD, FERTGOOD, SLOPMOD and PLOTDIS from home. The coefficients for the variables moderate soil fertility (FERTMOD) and good soil fertility (FERTGOOD) have the hypothesized negative sign and are significant at a 1% level. The associated odds ratio imply that when the soil fertility changes from poor soils to FERTGOOD the chances (odds) of a plot being abandoned decreases by about 14 times other factors remaining constant. Furthermore, the odd ratio (probability) of a plot remaining under abandonment decreases more (10 times) as the plot changes from poor soil to moderate soil status.

The variable SLOPMOD had the odd ratio of 2.09. This implies that the (odds) probability for the plot to be abandoned are greater when the plot is located on the moderate slopes (about 2 times) compared to the odds when the same plot is located on gentle slopes. Presence of stones (many) and plots on steep slopes are significant at 5% level. Abandoned plots are more likely to be observed on hilltops and mid slopes compared to plots in the valley bottoms. Cultivated plots are more common in the valley bottom and abandoned plots on top of hills. The presence of stones is also another factor that influenced plot abandonment. The odd ratio value is 1.52. This implies that the odds for the abandoned plots are greater (1.5 times) where there is presence of many stones compared with the odds ratio for it where the plots has less stones.

CONCLUSION AND POLICY IMPLICATIONS

The results reveal that plot abandonment and long fallow is a common problem in the Kigezi highlands. Firstly, long fallow and abandoned plots comprises over 25% of the land area. Secondly, while plot characteristics are important in the long run, farmers' attitude, social and economic factors contribute to farmers land use decisions. The findings indicate that household characteristics such as post primary education and households headed by females significantly contribute to farmers' decision to put land under abandonment and long fallow.

What can be done to address the issue of abandonment or low investment depends partly on the specific characteristics of plot. Analysis of the long fallows and abandoned plots show that 26% and 40% of each face multiple constraints to cultivation (e.g. steep slopes and stony). Just under half of each type face one constraint. The remainder, 30% of long fallows and 15% of abandoned plots appear to be highly cultivable. But even here, further analysis found that between 60 – 70% of the cultivable plots are located at least 500 meters from the homestead. This will impinge upon incentives for intensification and investment in valuable or labour intensive systems. For the cultivable plots and nearby plots, there are already available technological options and they can be put into production if resources permit. The large number of plots with physical constraints implies that on the technological side, options that require relatively little labour or management, such as woodlots or fodder production, need to be identified.

There is a further group of uncultivated plots that are far from the homesteads of the current owner. The study reveals that distance to fields hinders productive use of these plots. As a result, policy makers and farmers need to consult and find meaningful ways to encourage free exchange of land, renting and selling of far off plots. This is feasible in the Kigezi highlands because there already exists an active land market. But the same land market has helped to create the high levels of fragmentation on the landscape. The process of improving the spatial distribution of plot holdings will be slow so long as only bi-lateral transactions occur. Thus, the government could play a role in facilitating interested exchangers of land to meet and enable multiple party exchanges to take place.

This study has some limitations. For example, the farm level data on which the findings and recommendations are made are mainly for one season. Data covering more years would be better suited to meeting the objectives of this study. These data would relate more closely to farmers production practices. Because of limitations of time and money, this study focused mainly on Kabale district. While this can be extrapolated for Kigezi highlands, this may not be true for other highland areas in the country because of varying social-economic and climatic conditions.

REFERENCES

AFRENA (1999), *Quarterly Report*, NARO/ICRAF Project.

Artzen, J. W. (1984), Changes in Rural Activities and Utilization of Natural Resources in the Periods 1979-1983: The Case of Malolwane, Kgatleng, District, National Institute of Development Research and Documentation, University of Botswana: pp. 172.

Barbier, B. (1998), Induced Innovation and Land Degradation. Results from Bio-Economic Model of a Village in West Africa, *Agricultural Economics*, 19(1-2): 15-25.

Boserup, E. (1965), Conditions of Agricultural Growth, *The Economics of Agrarian Change under Population Pressure*. London, Allen and Uwin.

Djimbe, M., and Hoekstra, D. A. (1987), Agroforestry Potential for the Land Use Systems in the Bimodal Highlands of E. Africa. Nairobi, ICRAF.

Greene, W. H. (2003), *Econometric Analysis*, 2nd Edition, New York, Macmillan.

Grisley, W., and Mwesigwa, D, (1995), Socio Economic Determinants of Seasonal Fallowing: Smallholders in SW Uganda, *Journal of Environmental Management*, 42: 81-89.

Kashiwagi, M. (1998), Reconstruction of Regional Agriculture, and Farmland Resource Management in the less Favored Areas under the "Staple Food Law", *Journal of Rural Economics*, 69(2).

Lindblade, K. T. J., Carswell. G, Tumuhairwe J., (1998), Mitigating the Relationship Between Population Growth and Land Degradation. Landuse Change and Farm Management in SW Uganda, *Ambio*, 27(7).

Mmopelwa G, (1998), Factors Contributing to Land Fallowing in a Permanent Cultivation System: The Case of Semi Arid Botswana, *Journal of Arid Environment*, 40(2): 211-216.

MFPED (2004), Ministry of Finance, Planning, Economic and Development, Background to the Budget 2003/2004.

Zurayk R. A (1994), Rehabilitating the Ancient Terraced Lands of Lebanon, *Journal of Soil and Water Conservation*, 49(2): 106-112.

35

POPULATION DENSITY, SIZE OF OPERATION AND COLLECTIVE ACTION FOR MONITORING AND INVESTMENT IN RURAL INFRASTRUCTURE: IMPLICATIONS FOR POLICY

Patricia M. Makepe

INTRODUCTION

Rural agrarian communities in developing countries worldwide are primarily dependent on natural resources for their livelihoods. These communities are often faced with a plethora of constraints that adversely affect income generating activities and consequently impact negatively on quality of life. Specifically, when it comes to enterprise development, problems of inadequate infrastructure greatly inhibit the ability of enterprises to thrive; particularly where such ventures are located in areas that are remote and far away from cities. In Botswana cattle rearing is the mainstay of the rural economy. But the erratic nature of rainfall in the country means that there is a paucity of surface water, which is critical for cattle rearing. Consequently, the binding constraint for anyone intending to enter into the beef industry is the ability and capacity to gain access to a water point or borehole[1] that can provide a regular and sufficient supply of water for cattle[2] (White, 1993; Peters, 1994).

Developing and maintaining a borehole is very costly. While some individuals have managed to go it alone and own their boreholes as individuals, there are many instances where individuals have teamed together to form syndicates for purposes of pooling resources for developing and maintaining the borehole. Normally, individual members who are part of a syndicate are cash strapped and hence, there is an inherent incentive to act jointly or collectively to ensure that the borehole functions properly since without it their cattle will go without water. An added incentive for members to act collectively as a syndicate is that ownership of the borehole also gives the syndicate members *de facto* control over the grazing area surrounding the borehole within an 8km radius.

The government of Botswana has traditionally pursued policies that strongly support the growth of cattle production (Arntzen and Fidzani, 1997). These policies have come in the form of high prices for meat, input subsidies and a very lenient tax system. This strong support coupled with the utilization of borehole technology has led to over expansion of the cattle sector (Harvey and Lewis, 1990; Tsimako, 1991).

The literature on agricultural land use and location theory (Clark, 1973) predicts that near cities where population density is high and land is expensive, we should expect to find a large number of small sized operations[3] with low fixed costs and relatively good infrastructure. In contrast, far from cities where population density is low and land is cheap, we should expect to find a small number of large sized operations with high fixed costs and poor infrastructure. In addition, far from cities we may also find a small number of small sized operations facing high fixed costs and poor infrastructure. But on these small sized operations production is generally at a subsistence level.

Due to the factors outlined above, close to the city we should expect the need for investments in infrastructure to be lower than it is in areas far from the city because near the city most investments in infrastructure have already been made. In contrast, far from the city we should expect the need for investments in infrastructure to be higher because far from the city the infrastructure is relatively poorer.

Therefore, based on the population density and the size of operations found in each location, we should expect differences in the amount of collective action needed to make the required investment. Close to the city where there is high population density, a large number of small sized operations and little need for investment, collective action for making investments ought to be low and focused more on monitoring resource use. Far from the city with low population density, a small number of large sized operations and high need for investment, less collective action may needed to make the investment and monitor resource use because the small number of operations in this location makes self regulation easier. However, with a large number of small sized operations far from the city where there is a high need for investment, we expect more collective action focused on both making the much needed investments in infrastructure and also to monitor resource use.

The possibility exists that user groups can through collective action, team together to invest in and monitor resource development and resource use (Ostrom, 1990; Wade, 1988; Baland and Platteau, 1996; Hanna and Jentoft, 1996). By pooling resources their resources in this way, they can improve the availability of resource infrastructure critical for the survival of their enterprises and also enhance the performance of these enterprises; particularly in remote areas where such infrastructure is not well developed (Meinzen-Dick, *et al.*, 1996). As it relates to cattle production, syndicate members can pool their resources together and make investments in water provision and monitor resource use.

Previous research (Makepe, 2005) has established that while syndicate members do act collectively most of the time, there are instances in which members fail to cooperate with each other and this leads to problems with regard to the management of the resource. It is therefore, crucial to find ways in which collective action can be inculcated and made to function better to ensure the proper functioning of existing systems or institutions which can potentially improve rural livelihoods. In this way, the paper makes a contribution to the search for appropriate community based development strategies.

In this paper, collective action is taken to mean any action taken actively and jointly by syndicate members (or a committee of elected syndicate members) to achieve a common objective, whose outcome depends on the interdependence and reciprocity of syndicate members. Collective action occurs whenever members take steps actively to work together in managing water and grazing resources.

OBJECTIVES

The main objective of the study is to examine the relationship between population density, size of operation and collective action for investment and monitoring rural infrastructure. Based on the population density and the size of operations found in each location, it is expected that differences in the level of collective action needed to make the required investment will be observed. From these observations some policy recommendations will be prescribed which can effectively enhance collective action within the syndicates where it is necessary.

STUDY AREA

The study was conducted in Kgatleng district, located in southeastern Botswana. The district covers a total area of 7,960 km^2. It was selected for the study for the following reasons. First, all the grazing land in this district is under communal tenure and it is dominated by pastoralism. The district is relatively more densely populated compared to other districts in the country. As a result, the issue of fencing rangelands into ranches has been a polemic one so that the rangelands in the district remain under communal tenure. Finally, small and large cattle owners in the district have formed themselves into syndicates and drill boreholes to tap underground water, which provides permanent water for cattle production.

DATA

A survey of 73 borehole syndicates in Kgatleng District was conducted from February to September 2002. A list of all the boreholes that were drilled in Kgatleng district was obtained from the Department of Water Affairs in Gaborone, Botswana. From this list, all the boreholes that were owned by syndicates in the district were selected. Of the 80 syndicates, five declined to participate in the survey and so a total of 75 syndicates were interviewed. Of these syndicates, two were outliers[4] and were dropped from the analysis. As a result, data from the 73 remaining syndicates was used in the analysis.

The borehole syndicate was selected as the unit of analysis because it is the relevant institution responsible for borehole water provision and management. Therefore, it became expedient to look beyond the household as the unit of analysis. Boreholes are spread all over the northeastern part of the district, a predominantly grazing area covering more than 60 per cent of the district's total land area. The average number of cattle watered in one season per borehole is 1110. This figure consists mainly of members' cattle and a small number of in-migrating cattle, which make it to the borehole because the grazing area is not fenced.

Membership in a syndicate allows members to share ownership and control over the use of the borehole, and grazing resources within the vicinity of the borehole. The average number of members in a syndicate is 18. All members share joint responsibility for the borehole's operational and maintenance costs. These include the costs of fuel, oil, services and repairs. Tasks such as maintaining the troughs and fences surrounding the holding pens are done by members with assistance from herd boys (people hired to look after cattle). If a member's access to borehole water is terminated, other members do it as a sanction for his failure to pay his share of the borehole's operational and maintenance costs. This step is taken after several attempts at persuading the member concerned to pay his share of the operating and maintenance fees.

Each syndicate has a tiered structure, which extends access to the borehole beyond its members to include dependents and hirers. A member's dependents are the people who come behind him and gain access to the borehole through his membership to the syndicate. Dependents include the children or relatives of the member of the syndicate. For example, a son or a relative whose cattle are being held by the member qualifies as a dependent. On average, syndicate members allow four dependents access to the borehole. Dependents have no independent rights to the borehole except through the member and they do not pay fees. Nevertheless, non-monetary exchanges do occur between adult dependents and members, particularly if the dependent is not a child of the member. A dependent may offer his services as a herd boy to a member in return for access to borehole water and grazing. Therefore, while there may be no monetary exchange there is some reciprocal exchange-taking place between members and dependents over time.

The next categories of cattle owners who have access to the syndicate borehole are the hirers. Hirers are people who pay members a fee for the privilege of enjoying use rights to the borehole water and grazing. Each hirer pays an amount proportional his herd size. 34 per cent of syndicates have no hirers, while another 63 per cent have an average of two hirers but do not intend to accept any additional hirers because of resource scarcity[5]. Neither dependents nor hirers have any say in the running of the borehole.

A structured questionnaire was administered to groups of three individuals representing the members of each syndicate. The survey collected information about the structure and function of the syndicate; the use and maintenance of water and range resources, range degradation and improvements; causes and the extent of disputes over resource use; and market accessibility. In addition, informal interviews were carried out with key informants in the Ministry of Agriculture at the district level and community elders to provide supplemental information.

The survey data was explored to examine the hypothesized relationships outline earlier in the paper. Initially, the survey data were explored to determine how the borehole syndicates in the study area are actually distributed in terms of size of operation and population density. Average herd size per member was used to measure the size of operation and distance from cattle market was used to measure the population density. Since there were no data available on population density at the syndicate level for the study area, the distance traveled to cattle market was used as an indicator of population density. Distance traveled to the cattle market is a good indicator of population density because it is expected that more people tend to be settled near the market than far away from the market or it is expected that markets are located where there are more people.

The data were categorized according to the average herd size per member and the distance traveled to market based on the using two thresholds or cut off points, which emerged from the data. Different ways of categorizing the data using other possible cut off points were attempted. For example, first three groups that were classified into three different categories of size (small, medium and large) and distance (near, medium and far) were attempted in the hope of spotting a larger difference between the largest and smallest group, making the selection of the threshold value have less influence on the findings. But for one thing some cases some of the categories did not have any cases and for another thing there were no significant findings. In the final analysis, the criteria for selecting groups were based on the two actual cutoff points, which emerged from the data itself because this method gave the most distinct and identifiable groups.

For average herd size per member the cut off point used was 70 cattle. Syndicates with members having an average herd size of less than 70 cattle were taken to be small sized operations. Conversely, syndicates with members having an average herd size greater than 70 cattle were considered to be large sized operations. For distance traveled to the market the cut of point was 167 km. Syndicates that were located less than 167km from the market were considered to be close to the market and hence, were in high population density location. Conversely, syndicates located more than 167km from the market were considered far from the market and located in a low population density location. A scatter plot of these observations, which emerged from the data, can be seen in Figure 1.

Figure 1
A Scatter Plot of Borehole Syndicates According to Size of Operation and Distance to Market

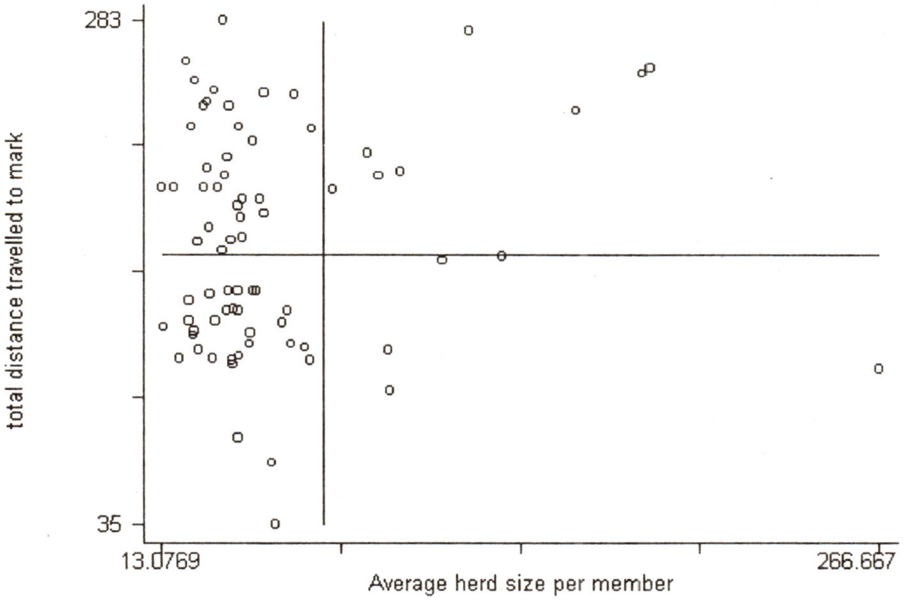

Source: Makepe (2002)

The above table is divided into four quadrants. Each quadrant represents one of the four groups. The quadrant on the bottom left hand side of the graph represents group one, small and near with less than 70 cattle and less than 167km from the market. The quadrant on the top left hand side of the graph represents group two, small and far with less than 70 cattle and more than 167 km from the market. The quadrant to the bottom right hand side represents group three, large and near with more than 70 cattle but less than 167 km away from the market. Finally, the top quadrant on the right hand side represents group four, large and far with more than 70 cattle per member and more than 167 km away from the market.

Group one consisted of 30 syndicates, which were small in terms of average herd size per member (< 70 cattle) and near the market (< 167 km). For this group, it was expected that a low level of investment and collective action focused on monitoring resource use would be found. Group two consisted of 30 syndicates, which were small in terms of average herd size per member (< 70 cattle) but were far from the market (> 167 km). For this group, it was expected that more investment accompanied by a higher level of collective action focused on both making the investment and monitoring resource use would be found. Group three consisted of 5 syndicates, which were large in average herd size per member (> 70 cattle) and near the market (< 167 km). For this group it was expected that a low level of investment and collective action would be found. Group four consisted of 8 syndicates, which were large in terms of average herd size per member (> 70 cattle) but far from the market. For this group it was expected that a high level of investment but with less collective action required to make that investment would be found.

For all four groups Analysis of Variance (ANOVA) was used to test the hypothesis that there are significant differences in the following variables of interest: investment made to improve the water source and the level of collective action across all four groups. Two indices related to collective action, which were developed earlier using factor analysis,[6] were used to measure the level of collective action.

RESULTS AND DISCUSSION

Previous research in which probit analysis was done to understand the determinants of collective action, Makepe (2005), showed that average herd size per member had a negative and significant effect on investments made to the water source. This was taken to mean that larger sized operations made fewer investments to the water source now because they had already made the investment earlier. The same probit analysis showed that the distance to the cattle market had a positive and significant effect on improvements made to the water source. This meant that those syndicates that were located far from the market made investments to improve their water source because they were located in areas where there is poor infrastructure. In that same analysis, wealth heterogeneity turned out to have a significant effect on investments made to the water source suggesting that in cases of significant wealth heterogeneity members with a larger stake in the investment would undertake the needed investment and collect money from other syndicate members later.

In light of these earlier findings (see Makepe, 2005) the ANOVA took the analysis further and explored the relationship between the four groups and wealth heterogeneity, investments made to improve the water source and collective action. The results from the ANOVA are shown in Table 1.

Table 1
Relationship between Average Herd Size Per Member, Total Distance from Cattle Market and Investments to Improve the Water Source and Level of Collective Action

Variable	*Small and near (n = 30)*	*Small and far (n = 30)*	*Large and near (n = 5)*	*Large and far (n = 8)*
Wealth heterogeneity	0.22	0.21	0.22	0.16
Between groups SS 0.03 F = 0.99 Prob. > F 0.40				
Improvements made to water source	0.30	0.80	0.47	0.50
Between groups SS 1.26 F = 1.76 Prob. > F 0.16				
Index of disputes, nonpayment and punishment	0.45	0.49	0.62	0.45
Between groups SS 0.51 F= 2.33 Prob. > F 0.08				
Index of number of meetings and percentage attendance	0.60	0.69	0.58	0.60
Between groups SS 0.06 F = 2.79 Prob. > F 0.04				

Source: Makepe (2002)

There are not very many significant relationships as can be seen from the lack of significant differences between wealth heterogeneity and investments made to improve the water source across the four groups. Nevertheless, the analysis on the two indices of collective action gave significant results. The index on disputes, nonpayment and punishment, which was chosen to represent members' not following rules, is almost the same across groups. This indicates that in the study area, syndicates face similar problems regarding members not following rules regardless of the size of operation and population density. For the index on number of meetings and percentage attendance at meetings chosen to represent members' active participation it is evident that for both small and large syndicates the level of collective action was significantly higher for syndicates that were farther away from the market. This confirms the hypothesis that farther away from the city we should expect more collective active. However, it was not possible within the limits of the data available to determine whether collective action was focused on making investments in the case of the small syndicates and on monitoring for the larger syndicates. Nevertheless, the results of the analysis show that syndicates in remote areas are more prone to working well together. Despite this observation there are some problems which hinder cooperation, which once adequately addressed can improve the performance of the syndicates in water provision and management.

CONCLUSION AND POLICY IMPLICATIONS

The data was examined in order to gain a deeper understanding of the relationship between population density, size of operation and collective action for monitoring and making investments in rural infrastructure. Specifically, ANOVA was used across the four categorized groups that emerged from the data and wealth heterogeneity; investments made to improve the water source and the level collective action. The analysis was done to clarify the reasons for observing more collective action for investment in resource improvements and monitoring resource use in remote areas.

It was found that in the study area, syndicates face similar problems which hinder their ability to work together collectively regardless of the size of operation and population density. For both small and large syndicates the level of collective action was significantly higher for syndicates that were farther away from the market confirming the hypothesis that farther away from the city we should expect more collective active. However, it was not possible within the limits of the data available to determine whether collective action was focused on making investments in the case of the small syndicates and on monitoring for the larger syndicates. Nevertheless, the results of the analysis show that syndicates in remote areas are more prone to working well together.

A resource that is to be managed or improved collectively must be small enough so that group members can effectively control and monitor the resource and exclude outsiders. Greater social cohesion within the group is also facilitated by a smaller number of users and higher net benefits derived from group membership. In remote areas, where there is lower population density and individuals tend to rely more on the resource for the livelihood collective action was much higher.

Despite this observation there are some problems which hinder cooperation, which once adequately addressed can improve the performance of the syndicates in water provision and management. The following conditions have been identified in the as being necessary conditions for user groups, in this case syndicates, to work well collectively in managing common pool resources (Runge, 1986; Uphoff, 1986; Wade, 1988; Ostrom, 1990; Rasmussen and Meinzen-Dick, 1995; White and Runge, 1995; Baland and Platteau, 1996). There must be a clear definition of members and boundaries of the resource to be managed and improved. There has to be a clear set of rules and obligations that are adapted to local conditions. Members must have the ability to adapt and modify the rules with changing circumstances. There must be an adequate and effective monitoring system in place. Sanctions for non-compliance must be enforceable and actually enforced. Furthermore, the syndicate must be recognized by the government authorities and not be undermined by them.

NOTES

1. A borehole is essentially a water point or a deep well, drilled in the ground to enable the cattle owner to utilize underground water for purposes of watering his cattle herd.
2. Prior to the establishment of the technology to mine underground water sources cattle rearing was restricted to those parts of the country with surface water. But since the advent of borehole technology in the 1950s cattle numbers have increased.
3. In this paper, the words operation and enterprise are used interchangeably in the context of the size of the cattle production activity.
4. In one syndicate there were only two members because all others had been evicted. Of the remaining members, one complained that the other member was not contributing to the running of the borehole at all. In fact, he was in the process of getting him evicted from the syndicate. The second syndicate had 100 members which was unusually high.
4. Increasing resource scarcity is evident in two ways: first, in the way syndicates turned down applications from cattle owners who wanted to be members of existing borehole syndicates, and second, in the practice of limiting the number of dependents the syndicate could have
4. For a detailed explanation of how these indices were constructed see Makepe (2005).
5. Increasing resource scarcity is evident in two ways: first, in the way syndicates turned down applications from cattle owners who wanted to be members of existing borehole syndicates, and second, in the practice of limiting the number of dependents the syndicate could have
6. For a detailed explanation of how these indices were constructed see Makepe (2005).

REFERENCES

Arntzen, J. W. and N. H. Fidzani (1997), Incentives for Sustainable Natural Resource Management and Economic Diversification in Botswana, Draft Report to the National Conservation Strategy Agency, University of Botswana.

Baland, J. M and J. P. Platteau (1996), *Halting Degradation of Natural Resources: Is There a Role for Communities*?, Oxford: Clarendon Press.

Clarke, C. (1973), *The Value of Agricultural Land*. Oxford Pergamon.

Hanna, S and S. Jentoft (1996), Human Use of the Natural Environment: An Overview of Social and Economic Dimensions, *Rights to Nature Ecological, Economic, Cultural and Political Principles of Institutions for the Environment*, Edited by S. Hanna; C. Folke and K. Maler, Island Press.

Harvey, C. and S. R. Lewis (1990), *Policy Choice and Development Performance in Botswana*, London: Macmillan in association with the OECD Development Centre.

Makepe, P. M. (2002), Three Essays on Collective Action and Communal Rangeland Management in Botswana, Unpublished Ph.D Dissertation for the Degree of Ph.D. Michigan State University.

Makepe, P. M. (2005), The Nature, Extent and Determinants of Collective Action over Water and Grazing Management in Botswana. *International Journal of Environment and Development*, (2): 1, June Issue.

Meinzen-Dick, R. S., L. A. Jackson and T. A. White (1996), Conceptual Framework: System Wide Program on Property Rights and Collective Action, International Food Policy Research Institute, Washington, D.C.

Ostrom, E. (1990), *Governing the Commons: the Evolution of Institutions for Collective Action*, Cambridge, U.K.: Cambridge University Press.

Peters, P. (1994), *Dividing the Commons: Politics, Policy, and Culture in Botswana*, Charlottesville and London: University Press to Virginia.

Ramussen, L. N. and R. Meinzen-Dick (1995), Local Organizations for Natural Resource Management: Lessons from Theoretical and Empirical Literature, Environment and Production Technology Division Discussion Paper No. 11, Washington D.C.: IFPRI.

Runge, C.F. (1986), Common Property and Collective Action in Economic Development, *World Development*, 14 (5): 623-635.

Tsimako, B. (1991), The Tribal Lands Grazing Policy Ranches Performance to Date, Agricultural Information and Public Relations Division, Gaborone, Botswana Uphoff, N. (1986), *Improving International Irrigation Management with Farmer Participation: Getting the Process Right*, Boulder, Colorado USA: Westview Press.

Wade, R. (1988), *Village Republics*, Cambridge, U.K.: Cambridge University Press.

White, R. (1993), *Livestock Development and Pastoral Production on Communal Rangeland in Botswana*, Food Production and Rural Development Division, Commonwealth Secretariat, London.

White, T. A. and C. F. Runge (1995), The Emergence and Evolution of Collective Action, Lessons from Cooperative Watershed management in Haiti, *Economic Development and Cultural Change*, 43 (1): 1-41.

36

LAND USE PATTERN AND ECOLOGICAL IMPLICATIONS: A COMPARATIVE STUDY OF UTTAR PRADESH AND MADHYA PRADESH

Gyan Prakash and Ram Kumar Jha

INTRODUCTION

Land resource forms the most important natural wealth of the nation and its proper utilization is a matter of concern. Its improper use leads to progressive deterioration and create ecological imbalances. The most important feature of land in our country is the large proportion of area available and suitable for agriculture. India is a very fast growing economy of the world. Tremendous transformation has taken place after introducing new economic policy. Industrialization and urbanization are sprawling at a faster rate. Thus land use pattern has also changed along with new facet of the economy.

The changes in land use pattern affect eco system in terms of vegetation, local climate, land quality, quality of life and water table. In other words, the changed in land use pattern is also associated with ecological change of area for example, to take higher yield, improper application of fertilizer, insecticide and pesticide have been used and these deteriorated the quality of land. The large proportion of agricultural land is suffering from either alkanity or salinity.

Land use statistics is an input for planner, policy maker and administrator to devise a comprehensive plan for short term as well as long term. The present land use pattern sketches the portrait of future because of its strong influence on land. Land use statistics are necessary for developing land use policies, land conservation activities and monitoring their effectiveness (Lindgren, 1985).

OBJECTIVE

An attempt has been made in this paper to examine the land use pattern and its ecological implications by using the land use changes in Uttar

Pradesh and Madhya Pradesh for the period from 1990-91 to 2001-02. For the analysis of data, the land use dynamic model used by Pandey and Tiwari (1987), Das (1989) and Pal (1991) has been adopted.

To examine the above objective the following information were required:

1. Total Reported area (R)
2. Forest(F)
3. Pasture and permanent grazing land (P)
4. Miscellaneous trees and groves (M)
5. Non-agricultural use (N)
6. Barren and unculturable land (B)
7. Culturable waste (W)
8. Net cultivated Area (C)
9. Current fallows (F_c)
10. Other fallow (F_o)

Since, the total geographical area/total reported area is constant, the changes 'if any' occurs/occur these changes will be inter class shift. These changes will be linearly additive over the time.

$$R = F + P + M + N + B + W^1 + C + F_C^2 + F_O^3 \quad (1)$$

The land use changes can be expressed as

$$\Delta R = \Delta F + \Delta P + \Delta M + \Delta N + \Delta B + \Delta W + \Delta C + \Delta F_C + \Delta F_O \quad (2)$$

Thus the rate of changes in land use will be equal to the rate of changes in the nine units. The accounting of the equation (2) does not facilitate to draw meaningful inferences about the direction and magnitude of the inter class shift. Hence, the total land endowment (i.e. reported area) has been divided into 4 broad categories

(a) Ecological Sector $E_1 = F + P + M$

(b) Ecological Sector $E_2 = B$

(c) Non-Agricultural Sector $N = N$

(d) Agricultural sector $A = C + W + F_o + F_c$

A prior knowledge of possible land use shift within and across the four sectors can be considered.

Possible favourable change for ecological balance

Land use if shifts

(i) M and P .... to .... F = It has no adverse effect on ecological implication

(ii) B .... to .... F = Favourable ecological implications

Possible adverse change for Ecological balance

(i) F .... to .... P = Up to some extent adverse impact on ecology

(ii) F and M .... to .... B = Serious adverse ecological effect

Keeping the above dynamics of land use, ecological sector has been divided into two sub sectors

(i) Desirable Ecology: $E_1 = F + M + P$

(ii) Undesirable Ecology: $E_2 = B$

Thus, net change in ecological sector ΔE can be written

$$\Delta E = \Delta E_1 + \Delta E_2$$

$$= (\Delta F + \Delta P + \Delta M) + (\Delta B) \quad (3)$$

As it is well known that there are no possibilities of land use shift from non-agricultural sector (N) to agricultural sector (A). So that net change in agricultural sector will have serious implication on ecology.

If net change is positive in agricultural sector ($+\Delta A$), it means it is at the cost of ecological sector.

If it is negative ($-\Delta A$), land use shift may be due to ecological or non-agricultural sector or both, but certainly at the cost of the agricultural sector.

Now, in agricultural sector (A) changes occur within agricultural sector, these will have some ecological implication.

$+\Delta A$ (a) = Leading to increase in net cultivated area ($+\Delta C$). It is favorable for agricultural growth.

(b) = There is no change in cultivated area ($\Delta C = 0$). It indicates serious and adverse phenomenon, like that of vicious circle, indicating a depletion of cultivated area, by the way of land use shift to Fc, Fo and W or one hand and depletion of area under F to keep the food production. This phenomenon will be the worst because of two reasons, first it will be the cost of the ecological sector and second it will be never favourable to agricultural growth.

Now, the overall inter-sect oral land use shift will be considered as:

$$\Delta R = \Delta E_1 + \Delta E_2 + \Delta A + \Delta N \quad (4)$$

Eq. (4) is nothing; it is same as equation (2). It has been expressed only in sectoral format to judge inter-sectoral shift.

DATA BASE AND METHODOLOGY

The data for calculation of land use shift have been compiled from the various publications of Government of Uttar Pradesh and Madhya Pradesh. The required land use statistics has been compiled from Government of Madhya Pradesh compendium of Agricultural statistics

2003-2004, published by M.P. State Agricultural Marketing Board, Bhopal. The basic land use statistics in M.P published by the State Directorate of Agriculture (Agricultural Statistics Division) has been used for the estimation.

Uttar Pradesh and Madhya Pradesh both are not only the largest States of the country but centrally located also. Both have common nature of agro-climatic features and have good natural resources but both represent BIMARU states. In other words, these are poor and have low per capita income. These are the reasons to make a comparative analysis of land use pattern and its impact on ecology.

To estimate annual rate of change and over the period of change linear and log liner models have been used respectively.

Linear Model

$$Y = a + b * t + U_t \tag{5}$$

Where,

$Y = (F/P/M/B/W/C/F_c$ and $F_o)$ i.e. 8 linear model separately.

t = time

a = intercept

b = coefficient (rate of change)

U_t = error term

Further the annual growth rate in the linear trend has been defined as

$$\text{Annual GrowthRate} = \frac{b}{Y} \times 100 \text{ (in percent)}$$

Where,

Y = Average of Y

Log linear model

$$Ln\ Y_t = \beta_1 + \beta_2 t + U_t \tag{6}$$

Where,

$Y = (F/P/M/B/W/C/F_c$ and $F_o)$ i.e. 8 log linear model separately.

t = time period

β_1 = intercept

β_2 = instantaneous growth rate

U_t = error term

To work out over the period of change i.e. Compound growth rate (C.G.R.)

$$= [\text{Antilog } (\hat{\beta}_2) - 1] \times 100.$$

RESULTS AND DISCUSSION

The summary of the results is given in Table 1. The log linaer trend equation gives better results hence it has been applied for estimation. The C.G.R. in Uttar Pradesh and Madhya Pradesh have recorded growth rate 1.69 per cent and 0.87 per cent respectively. This growth can be due to industrialization and urbanization. In ecological sector, forest area in Madhya Pradesh is increasing at a faster rate i.e. 1.046 per cent rate of growth than Uttar Pradesh. In U.P. it has been observed very slower in comparison to Madhya Pradesh. It is a serious matter of concern in Uttar Pradesh, the permanent posture and grazing lands are reducing at the rate of 3.88 per cent and 0.4618 per cent respectively in Madhya Pradesh and Uttar Pradesh. The depletion rate of miscellaneous trees and groves has been recorded highest in Madhya Pradesh i.e. 6.65% which is statistically significant also at 5% level.

Table 1

Compound Growth Rates and Annual Rate of Change of Various Land Use Classes in Uttar Pradesh, Madhya Pradesh (1990-91 to 2001-02)

Land Use Classes	*Uttar Pradesh*		*Madhya Pradesh*	
	C.G.R.	*A.R.C.*	*C.G.R.*	*A.R.C.*
Non-agricultural uses (*N*)	0.8723*	0.3754*	1.6986*	0.7319*
Forests (*F*)	0.3046**	0.1322**	1.0467*	0.4493*
Pastures and grazing lands (*P*)	–.4618**	–.2026**	–3.8897*	–1.6807*
Misc. trees and groves (*M*)	-0.2530	–0.1095	–6.6466**	–2.9760**
Barren and uncultivable lands (*B*)	–2.4921*	–1.0985*	–4.3863*	-2.0083*
Culturable wastes (*W*)	–3.5305*	–1.5670*	–1.0103	–0.4525
Net cultivated area (C)	0.5141*	0.2226*	0.2379	0.1030
Current fallows (F_c)	–2.4202*	–1.0668*	0.7477	0.4989
Other fallows (F_o)	–4.7796*	–2.1067*	–1.3630	–0.5803

*Significant at 1% level of significance
**Significant at 5% level of significance

The computation (Table: 1) shows that barren land is depleting at a faster rate in Madhya Pradesh i.e. 4.38 per cent. In Uttar Pradesh the current fallow land is decreasing but in Madhya Pradesh it is increasing. In Uttar Pradesh and Madhya Pradesh, other fallow land has been observed declining. The net cultivated area is increasing at the rate of 0.51 per cent in Uttar Pradesh and at 0.23 per cent in Madhya Pradesh.

The annual rate of change has been computed from Table: 1 to get the picture of inter- sectoral as well as annual rate of change in ΔN, ΔE_1, ΔE_2 and ΔA.

From the Table 2 the following facts can be derived:

The substantial land shifts have taken place from the undesirable ecological sector E_2 to other sectors across the states. This trend is highly appreciable, favourable and positive but the depletion of desirable ecological sector E_1 has been observed across the states Uttar Pradesh and Madhya Pradesh. It is a serious concern for economic growth. It seems, a part of this sectoral land use has shifted towards the non-agricultural sector due to urbanization and industrialization besides these some part of lands have shifted for the development of rural infrastructure.

Table 2
Sectoral Land Use Dynamics

Regions	*Annual rate of change*			
	Non-agricultural sector	*Ecological sector*		*Agricultural sector*
	DN	*D E_1*	*DE_2*	*DA*
Uttar Pradesh	0.3754	–0.1798	–1.0985	–4.5179
Madhya Pradesh	0.7319	–4.2074	–2.0083	–0.4309

Table 3
Summary of Comparison

Variables	*Uttar Pradesh*	*Madhya Pradesh*
Rainfall	+	–
Canals	–	–
Tanks	–	–
Tube wells	+	+
Growth of cities	+	+
Net Irrigated area	+	+
Rural (decadal gw)*	+	–
Urban (decadal gw)*	–	–

*Still urban population growth rate is higher than rural

Further attempt has been made to do comparison on the basis of two point of time i.e. 1990-91 and 2000-01 between M.P. and U.P. in the light of following variables like rainfall, numbers of canals, tanks, tube wells decadal growth rate of rural and urban population, growth of cities (Urbanization) and net irrigated area etc which may be the cause of ecological degradation.

In both states the similar trend has been observed. Now, there is a question whether these states are in vicious circle of ecological degradation or not. It is well known that rainfall depends upon the nature of the forest. More the spread of forest area is more the rainfall. Now as per scientific logic these states are in vicious circle of ecological degradation. (Analysis as per Table: 2 and 3)

As $E_{1\,(F,\,P,\,M)}$ area declines

Rainfall declines

It reduces availability of water on surface

More and more water will be extracted from underground
(Mechanical and electrical energy will be required)
Numbers of pump set and tube wells will be used

Why more water?

- To meet the requirement of food grains due to increase in population (Use of HYV of seeds its prerequisite condition is water)
- For Domestic Purposes (Human being and animals)
- For industrial purposes

These lead to lowering water table and shortage of water affects flora and fauna

Because soil looses humidity and moisture
Ultimately E_1 (F, P, M) also suffers from scarcity of water

Its (E_1) area reduces due to lack of water

It reduces once again rainfall
And
The process goes on

POLICY IMPLICATIONS

The question arises how and what type of strategy for economic development and growth should be evolved in the light of present analysis of land use so that economic growth process may make least harm to environment. It is well known that as economic growth process is on, some trade off between development and environment degradation will start. For e.g., if government tries to raise the production and productivity of agricultural produce to meet the requirement of increased population, this action will affect the local eco-system which consists of vegetation, flora and fauna, forest etc. As table 2 shows, the agricultural sector is depleting across the state level, but the question is, if it is running in the same fashion, what would be of agricultural growth and environment. More serious concern is about the depletion of ecological sector E_1, which consists of forest, miscellaneous trees groves and permanent pastures lands. Suppose E_1 deteriorates at present rate, no doubt, its consequences will be very horrible in future.

SUGGESTIONS

The following suggestions needs immediate attention by appropriate authorities:

1. *The government should make a micro level comprehensive planning to handle land use properly.
2. Extension of urbanization should be checked.
3. Bio-manure should be encouraged instead of chemical fertilizer.
4. The site of industrial estate should be assigned to those areas which are Barren or unculturable.
5. The constitutional amendment should be made to include the responsibilities of protection; conservation and regeneration of forest and the responsibilities should be shoulder on rural peoples, Gaon Sabha and Gaon Panchayat to check deforestation at the gross root level.
6. Commercial forestation like medicinal plants and Bio-fuel plants like Jetrofa should be encouraged. It will on one hand give additional revenue to government and income to people and forest area will be increased on the other hand.

ACKNOWLEDGEMENT

The authors are grateful to Prof. Ganesh Kawadia, Head, School of Economics, Devi Ahilya University, Indore for his valuable suggestions.

NOTES

1. Culturable Waste :- indicates those land lying fallow beyond 5 years.
2. Current fallows :- means those lands lying fallow for less than one year.
3. Other fallow :- shows land lying fallow between one and five years.

REFERENCES

Das, H. C. L. (1989), Land Use Dynamics and It's Ecological Implications in India: An Inter-Regional Analysis (1966-67–1982-83), 22nd Annual Conference of Regional Science Association, Bhopal.

Lindgren, David T. (1985), Land Use Planning and Remote Sensing, Mastinus Nijhoff Publishers, Boston.

Pandey, V. K. and S. K. Tiwari, (1987), Some Ecological Implications of Land Use Dynamics in Uttar Pradesh, *Indian Journal of Agricultural Economics*, (XLII)3: 388–394.

Pal Saroj K. (1991), Land Use Dynamics and It's Related Aspects in India: A Study in Time Series Analysis, 1950-84: in *Environmental Degradation and Developmental Strategies in India*, Ashish Publishing House, New Delhi, 280-302.

INDEX